João Jones da Silva
Luciana Rocha de Lima

Phytoremediation and ploughing in soils affected by salts

João Jones da Silva
Luciana Rocha de Lima

Phytoremediation and ploughing in soils affected by salts

Alternative method for recovering soils with salt problems

ScienciaScripts

Imprint

Any brand names and product names mentioned in this book are subject to trademark, brand or patent protection and are trademarks or registered trademarks of their respective holders. The use of brand names, product names, common names, trade names, product descriptions etc. even without a particular marking in this work is in no way to be construed to mean that such names may be regarded as unrestricted in respect of trademark and brand protection legislation and could thus be used by anyone.

Cover image: www.ingimage.com

This book is a translation from the original published under ISBN 978-613-9-70235-0.

Publisher:
Sciencia Scripts
is a trademark of
Dodo Books Indian Ocean Ltd. and OmniScriptum S.R.L publishing group

120 High Road, East Finchley, London, N2 9ED, United Kingdom
Str. Armeneasca 28/1, office 1, Chisinau MD-2012, Republic of Moldova, Europe
Printed at: see last page
ISBN: 978-620-8-17488-0

"The best learning is to think that the value of things is not in the time they last but in the intensity with which they happen." (Fernando Pessoa)

To God, the almighty father, without whom we wouldn't be able to get anywhere, who is the source of love and of life to the full.

To my parents, José Pedro da Silva and Valderice Maria de Andrade (in memoria), for their affection, tolerance, love and encouragement.

To my dear wife, Luciana Rocha de Lima, for her love and affection, always being by my side, supporting me and helping me to overcome another stage of life.

To my dear little son João Lucas, whom he inspires and cherishes.

DEDICATE

ACKNOWLEDGEMENTS

To God, who gives us the strength and courage to face all our challenges;

To my advisor, Professor Manoel Moises Ferreira de Queiroz, to the other professors, without wishing to belittle the others, they all contributed in one way or another, but I would like to mention with great care my professors, Reginaldo Gomes Nobre and Marcos Eric Barbosa Brito, who undoubtedly made a special contribution to this work.

Professor Oscar Mariano Hafle from the IFPB also played a special role in the production of the seedlings for the experiment, and Professor Ednaldo Barbosa Pereira Junior, also from the IFPB, immediately agreed to take part in the panel as an external examiner, and his contribution was also indispensable.

I would like to thank my colleague, Professor Hermano Oliveira Rolim, an agronomist at the IFPB, for his help in conducting the experiment, along with me in all the phases of the experiment, in the laboratory analyses, his help was sensational, without his help Hermano, I would run a lot, [laughs], so my thanks to you Hermano, and finally, to everyone, without distinction, to all the professors at the UFCG Postgraduate Course, Pombal campus, for their dedication and for my learning, I thank you.

I also couldn't forget that sensational person, Kadydja Mayara Ramos Nobre, secretary of the Postgraduate Programme at the Pombal Campus, your attention to us was always sensational, you always answered with that cheerful smile, you never failed to answer in any other way, always smiling and attentive, my thanks to you Kadydja, your attention, I can never forget it. Finally, to everyone who contributed in one way or another, my sincerest THANKS, THANK YOU!

SUMMARY

SUMMARY

The Northeast region of Brazil is characterised by a semi-arid climate, with rainfall volumes normally lower than evaporation, making the use of irrigation essential for sustainable agricultural production. However, inadequate irrigation management linked to local conditions has favoured soil salinisation and sodification processes, leading to the degradation and abandonment of large areas that were previously productive. Among the areas affected by salts in this region, the São Gonçalo Irrigated Perimeter - PB stands out for its degraded soils with salt problems, which prompted the proposal to study methods and techniques aimed at improving the soil's physical and chemical characteristics. The experiment was set up in an area previously identified as having saline-sodic soil in the São Gonçalo Irrigated Perimeter, and to recover it, the methods and technique of phytoremediation were applied using salt grass (Atriplex numulária L.), chickweed (Eleusine indica L.), wild parsley (Ipomoea asarifolia L.) and agricultural gypsum. The research consisted of eight treatments, with four replications. The treatments were subjected to a randomised block design (RBL), with the following treatments: (i) cultivation of atriplex; (ii) cultivation of crowfoot grass; (iii) cultivation of parsley; (iv) soil with gypsum; (v) application of agricultural gypsum associated with the cultivation of atriplex; (vi) application of agricultural gypsum associated with the cultivation of crowfoot grass; (vii) application of agricultural gypsum associated with the cultivation of parsley and (viii) soil with no management (control), in a total area of 62 m^2 conducted in the field, in plots measuring 1.40 x 1.40 m. The aim of this study is to recover soil affected by salts using an alternative, more economically viable and sustainable method using plants capable of extracting excess salts from the soil.

Keywords: Atriplex numulária L., Eleusine indica L., Ipomoea asarifolia L., sodification, soil recovery.

CHAPTER 1

INTRODUCTION

Plant production is the main link between the soil and its constituents and the other players in various ecosystems. Based on this, it is up to soil scientists and the community in general to seek alternatives that minimise the impacts of anthropogenic causes, which most often result in a reduction in soil quality. When this damage directly affects the possibility of cultivation, linked to the maximisation of surface runoff and soil loss, reduced infiltration and consequent recharge of aquifers, among other causes, it is necessary to take effective action (SOUZA, 2010).

So, although soil salinity is a recurring issue, especially in the semi-arid northeast of Brazil, it has not lost its current focus, since it reduces productive capacity in a country with such a great need for food, especially in the places where this problem is most prevalent. For its recovery and/or remediation, more costly alternatives such as the use of chemical correctives and machinery to set up drainage systems are not always available. Therefore, there is nothing more convenient than the use of vegetation that can withstand high saline levels, maintain them in its plant tissues, produce large amounts of biomass and provide food for certain animal species. This gives rise to the term phytoremediation and, more specifically, phytoextraction (SOUZA, 2010).

Recovering degraded areas aims to provide the degraded environment with favourable conditions for the restructuring of life in an environment that does not have the physical, chemical and/or biological conditions to regenerate on its own. The recovery of a degraded area can be achieved through correct soil management, an efficient drainage system or the planting of plant species (SDB, 2015).

The supply of salts by irrigation water or their solubilisation from the mineral components of rocks can lead to an accumulation of cations and anions that compromise the productive capacity of soils, making them unsuitable for agriculture. Thus, soil salinity and sodicity have become globally recognised environmental problems, degrading not only the areas causing the degradation, but also neighbouring areas that receive the contribution of salts released into the environment (MIRANDA, 2013).

There are countless factors that can contribute to the formation of soils affected by salts, provided that all these factors are always related to climatic conditions where evapotranspiration rates exceed precipitation rates (a typical condition in arid and semi-arid climate regions), as well as a poor drainage system. Some soils, under these conditions, tend to be more prone to the phenomenon of salinisation, and this greater or lesser susceptibility on the part of these soils will depend heavily on their physical, chemical and mineralogical characteristics (MIRANDA, 2013).

Areas affected by salts occur over large areas of the world, mainly in regions with arid and semi-arid climates, where irrigation is fundamental for the development of successful agriculture. The increase in the content of soluble salts in the soil influences crop behaviour in various ways, through changes in the proportions of exchangeable Na^+ , the reaction of soils, their physical properties, the osmotic potential of the soil solution and the toxic effects of specific ions. These changes influence the activity of plant roots and soil microorganisms, and consequently crop productivity (CAVALCANTE et al., 2010).

These problems are intensifying, especially in irrigation projects, due to inadequate management of the technique, which is reflected in reduced crop productivity. According to the FAO (2000), the total area of soils affected by salts in the world, including saline and sodic soils, is 831 million hectares, of which 397 million are saline soils and 434 million are sodic soils.

With regard to the problem of salinity in Brazil, the Sertão region of the Northeast stands out. However, other areas are also affected, such as some places in the Amazon region and in the north of Minas Gerais. However, the largest proportion of salinised soils in Brazil is found in the Northeast, due to its climatic conditions. In this region of Brazil, the area exploited with irrigation is still small, totalling approximately 663,672 ha, but there is potential to reach 1,304,000 ha (CHRISTOFIDIS, 2001).

As irrigation is seen as one of the alternatives for the socio-economic development of semi-arid regions, it must be rationally managed in order to avoid problems of excess salts in the soil and the degradation of water and soil resources, since the climatic conditions of these regions are favourable for the occurrence of soil salinisation (MEDEIROS et al., 2010).

It is therefore necessary to know the areas degraded by soil salinity and sodicity, the causes that triggered the degradation processes and to develop more efficient techniques suited to the conditions of north-eastern Brazil, making it possible to minimise the progress of degradation and improve soil properties for the cultivation of crops of agricultural interest.

The aim of this study was to recover areas affected by salts in the São Gonçalo Irrigated Perimeter, in the Sertão region of Paraíba, through the use of alternative management of soils affected by salts by phytoextraction with Atriplex and other invasive plants typical of this region and the use of agricultural gypsum in an attempt to recover soils with salt problems using low-cost techniques and methods.

CHAPTER 2

LITERATURE REVIEW

2.1. SOILS WITH SALT PROBLEMS IN IRRIGATED PERIMETERS

Soil salinisation is a growing problem worldwide, with billions of hectares of soil believed to be altered by salts, and a large part of all irrigated areas in the world suffering from reduced production due to excess salts (KEIFFER & UNGAR, 2002; HORNEY et al., 2005).

With the rapid growth of the world's population, the demand for food is constantly increasing. And to meet the needs of this growing consumption, the agricultural sector is increasingly driving food production. As a result, irrigation has become a very attractive tool, as it has made it possible to increase arable areas all over the world, including making arid and semi-arid regions productive (MEDEIROS et al., 2010).

The regions with arid and semi-arid climates, which include the water-scarce Northeast of Brazil, covering 150 million hectares, have favourable conditions for soils affected by excess salts, due to the characteristics of the climate, relief, geology and drainage, among other factors. In this part of the Northeast, most of the irrigation perimeters are affected by degradation, ranging from a reduction in crop yields to the abandonment of the areas (MOTA & OLIVEIRA, 1999; BARROS et al., 2004; RIBEIRO et al., 2010).

Environments with arid and semi-arid climates, characterised by low rainfall and high rates of potential evapotranspiration, and fragile due to the limited water reserves that sustain animal and plant life, these regions are more likely to have areas with salt problems. To enable agricultural production in these environments around the world, the technique of irrigation has been disseminated, based on the principle of supplying the water needed for plant production, which would then provide food for animal and human consumption. However, this technique, when poorly managed, can lead to immeasurable environmental problems, such as the degradation of soil and water resources, making them unsuitable for agricultural use.

With the increase in irrigation in recent decades, especially in the north-eastern region of Brazil, as it is a region of great irregularities in the occurrence and distribution of rainfall. However, there has been a greater tendency to use irrigation systems and build irrigated perimeters, often with inadequate management for local conditions, and the processes of soil salinisation and sodification have intensified, leading to the degradation and abandonment of large areas that were previously productive. Let's take a look at the number of Public Irrigated Perimeters in Brazil with their respective irrigated area (Table 1).

Table 1. Irrigated Areas in Public Perimeters in Brazil

Irrigated Area in Public Perimeters< Responsible Body	Quantity	Irrigable Area (ha)
São Francisco and Parnaíba Valley Development Company	35	362.053,00
National Department of Drought Works	31	70.227,80
Ministry of National Integration	01	62.000,00
Tocantins State Agriculture and Livestock Secretariat	02	10.129,00
Sergipe State Secretariat for Agriculture and Rural Development	04	16.808,00
State Secretariat for Infrastructure/RR	01	1.000,00
Paraíba State Secretariat for Agricultural and Fisheries Development	01	6.335,70
Federal University of Pelotas/Mirim Lagoon Agency	01	25.000,00
TOTAL	76	553.553,50

Source: SENIR (2015)

There are several examples of salinisation processes in irrigated perimeters as a result of inadequate irrigation, as cited by Ribeiro et al. (2003).

Also according to authors such as Batista et al. (1998), salinisation resulting from irrigation has been observed in regions with annual rainfall of up to 1,000 mm, mainly in shallow soils with poor drainage. In addition to low rainfall, the high water deficit resulting from high evapotranspiration rates contributes to the process of salt accumulation. This water deficit, according to Ribeiro et al. (2003), prevents rainfall from promoting complete and frequent washing of irrigated soil profiles, so that the ions Ca^{2+}, Mg^{2+}, Na^+, K^+, Cl^-, $SO4^{2-}$, $CO3^{2-}$ and HCO^{3-} in solution tend to remain in the root zone.

The appearance of salinity in irrigated areas is common, caused by irrigation water containing high concentrations of salts, as a result of management practices that do not aim to preserve the productive capacity of the soil, the absence of a drainage system, the quantity of water and the indiscriminate and excessive use of fertilisers (BERNARDO et al., 2006).

In the north-eastern region of Brazil, the largest area affected by excess salts in the soil is located in the west of Bahia (44% of the total), followed by the state of Ceará, with 25% of the total area (FAGERIA & GHEYI, 1997). In these states, salinity has been identified as one of the main factors responsible for the reduction in crop growth and productivity (PEREIRA et al., 2005).

Saline soils have a CEES of more than 4.0 dS.m^{-1} (deci-siemens per metre) and can be recovered by leaching the salts (BERNARDO, 1995). On the other hand, sodic soils have a PST of more than 15 per cent and can be improved by applying conditioners (BERNARDO, 1995) or by subsoiling (HOLANDA, 2000).

There are several reports of soils with salt problems in various irrigated perimeters located in the northeastern states. Some of these sites affected by salts are cited by Macêdo (1988) in terms of percentage of area: Custódia-PE with 97%; Ceraíma-BA with 32%, and Cachoeira II; PB with 30%. Suassuna & Audry (1993) described the percentage of irrigated areas with salinisation problems in these regions as approximately 32%, which could increase if preventative measures are not adopted. Aguiar Netto et al. (2006) indicated salinisation problems in the irrigated perimeters of Bebedouro and Nilo Coelho - PE, Tourão - BA, Morada Nova and Curu-Paraipava - CE and Jabiberi - SE, where 76.5% of the plots studied were saline-sodified.

The harmful effects of salinity are toxicity caused by high concentrations of chemical elements in the soil such as Na$^+$, Ca^{2+} , Mg^{2+} and Cl$^-$; a decrease in the osmotic potential of the water in the soil, which makes it difficult for crops to absorb water; and loss of soil structure, specifically when there is an increase in the percentage of exchangeable sodium (PST) above 15 per cent (USSL STAFF, 1954).

Therefore, in the north-east of Brazil, where the irrigated perimeters are located, combined with edaphoclimatic factors and the use of water with high levels of salts, as well as the inadequate use of irrigation and an inefficient drainage system, there has been an increase in areas with salt problems.

2.2 . SOILS WITH SALT PROBLEMS

The problem of soils affected by salts involves three main and sequential processes that lead to the development of these soils: salinisation which gives rise to saline soils, solonisation which promotes the formation of sodic soils and consists of two sub-processes. The first, sodification, is the process whereby the Na+ ion passes from the soil solution to the exchange complex, forming the so-called saline-sodic soils. The second, desalination, washes away the soluble salts, resulting in the formation of only sodic soils. Finally, there is the possibility of a degradation process called solodisation, which leads to the washing away of the Na+ ion and its replacement by the H+ ion, giving rise to non-saline and non-sodic soils (RIBEIRO et al., 2009).

The process of soil degradation worsens when the Na+ ion predominates over calcium and magnesium salts, making soil management more problematic, as changes occur to the soil's physical properties.

In these soils, characterised as sodic and saline-sodic, sodium becomes a high proportion of the exchange complex. This establishes a dispersion condition in the soil, in which the dispersed colloids are moved in the profile, clogging pores in subsurface horizons and thus altering the soil structure (RUIZ et al., 2004; QADIR et al., 2007; DIAS & BLANCO, 2010).

In order to understand the classification of soils affected by salts, some chemical properties of the soil are used, such as: hydrogen potential (pH), electrical conductivity of the saturation extract (ECes) and Percentage of Exchangeable Sodium (PST) (Ribeiro et al., 2003).

The difference between halomorphic and non-halomorphic soils is established, according to the United States Salinity LINITY LABORATORY - USSL Staff (1954), by specific values for the properties that classify them (Table 2).

Table 2. Classification of soils affected by salts (Richards, 1954).

Classification	ECes (dSm-1)	pHps	PST (%)
Saline	>4	< 8,5	< 15
Saline-sodic	>4	< 8,5	>15
Sodium	< 4	> 8,5	>15
Normal	<4	<8,5	<15

Soil salinity is measured by the electrical conductivity in the saturation extract (ECes), while sodicity is assessed by the percentage of exchangeable sodium (PST) in the soil's cation exchange complex. EC values above 4 dS m^{-1} and PST values above 15% characterise the soil as saline-sodic, United States, (1990). Knowing that there are studies that recognise the classification considering EC less than 2 dS m^{-1} , as most sensitive plants would not be able to withstand an EC of 4 dS m .$^{-1}$

Therefore, according to this classification, we can differentiate saline, saline-sodic and sodic soils, with ECes, pH and PST.

2.3 TECHNIQUES FOR RECOVERING SALINE-SODIC SOILS

The most appropriate management practices for controlling soil salinity in the long term are fundamental in a programme for cultivating soils affected by salts, especially with a view to sustainability, as well as being alternatives for soil use and recovery. Among the techniques for recovering saline-sodic soils, the application of chemical correctives and soil washing are widely used, as they act directly to correct the problems of these soils in relation to plants (QADIR et al., 2001; MONTENEGRO & MONTENEGRO, 2004).

However, phytoextraction of soluble salts using halophytic plants is a low-cost alternative for recovering saline soils that are not aggressive to the environment. According to Qadir et al. (2007), phytoremediation is an efficient strategy for recovering saline-sodic soils, with performance comparable to the use of chemical correctives. Qadir et al. (2001) concluded that phytoextraction shows recovery effects comparable to those of applying gypsum, farmyard manure or irrigation water treated with sulphuric acid. They also observed that phytoremediation had lower implementation and management costs than treatments with chemical correctives.

The soils were classified according to sodicity using the criteria proposed by Massoud (1971), quoted by Pizarro Cabello (1985; 1996), which group PST classes according to (Table 3).

Table 3. Classification of soils according to sodicity as a function of PST.

Class	PST
Non-sodic	< 7%
Slightly sodium	7% - 10%
Medium sodium	11% - 20%
Strongly sodic	21% - 30%
Excessively sodium	> 30%

However, there are other techniques that can be used to correct saline-sodic and sodic soils: in addition to agricultural gypsum, sulphur, aluminium sulphate, calcium chloride and sulphuric acid can also be used. Due to the quantities found in the Fertiliser Industries, as a by-product in the manufacture of simple superphosphate, gypsum is the most suitable because of its lower cost.

For the recovery of saline soils, washing is the most suitable method. According to Pizarro (1985), quoted by Tavares Filho (2010), salts should be leached from saline soils in two ways: continuous washing, which covers the surface of the soil with a 10 cm high sheet of water. Here, the salts are removed more quickly and allow the area to be worked sooner. This method is recommended for soils with good permeability, a deep water table and a high evaporation rate.

Intermittent flushing is recommended for soils with low drainage capacity, a high water table and low salinity groundwater. It should be applied during periods of low evaporation. The salts should be leached out in the cold season when evaporation is lower. Cultivate species that are more tolerant to salinity, as there is a saving in water due to less leaching. Avoid prolonged periods without irrigation, as this favours the accumulation of salts.

2.4 SALINITY TOLERANCE

Some plants tolerate soil with excess salts well, while others do not. Therefore, the concept of salinity tolerance is based on the ability of plants to develop under conditions of saline stress (MUNNS, 2002). In order to adapt to the adverse conditions generated by the high levels of salts in the system, halophyte plants have certain survival mechanisms (ZHU, 2001).

Plants that have the ability to survive in adverse conditions with resourcefulness, for example, in reducing osmotic potential (ARAÚJO et al., 2006), thus managing to overcome the effects of water restriction and the presence of high concentrations of salts and sodium in the environment. This group of plants that respond positively to salinity is classified as halophytes (KOYRO, 2006).

2.5 EFFECTS OF SALTS ON PLANTS

Salts have various effects on plants, with some species being more tolerant and others more sensitive. Responses vary according to the levels and types of salts, as well as between and within plant species.

One of the effects of high concentrations of salts in the soil is osmotic stress. This is the most striking effect caused by salinity on plants, which restricts the absorption of water by plants due to the decrease in osmotic potential in the soil solution. According to the laws of thermodynamics, the energy potential of a substance increases as a function of its concentration. And this substance will move from areas with higher potentials to those with lower potentials (EPSTEIN & BLOOM, 2006). The concentration of salts can reach such high values that the plant will eventually lose water to the external environment (DIAS & BLANCO, 2010).

In addition to these osmotic effects, when the plant absorbs saline ions, plant toxicity can occur due to an excess of absorbed salts. These effects interfere with plant growth and development, affecting the normal functioning of enzymatic activity and, consequently, physiological and biochemical processes such as respiration, photosynthesis, protein synthesis and lipid metabolism (ESTEVES & SUZUKI, 2008; FREIRE & RODRIGUES, 2009; DIAS & BLANCO, 2010; GONÇALVES et al., 2011). In addition to these effects acting directly on plant development, Barros et al. (2009) highlight indirect reactions arising from the effects of salts on the soil. The structure of sodic soils prevents seed germination and root development, as well as increasing the plant's energy consumption, posing serious problems for agricultural productivity. Therefore, salts cause a series of problems for plants, interfering with the entire metabolic system, affecting the physiological system and consequently reducing production.

2.6 PHYTOREMEDIATION

Phytoremediation is a low-cost technique for recovering soils. Some research has pointed to phytoremediation as an alternative for managing areas degraded by salinisation and sodification, by growing plants capable of accumulating excess salts and removing them from the soil. To do this, it is necessary to use plant species that can not only tolerate high levels of salts, but also produce enough biomass to extract considerable quantities of salts (FREIRE et al., 2010).

In order for salt phytoextraction to be successful in saline-sodic soils, the plants must show tolerance to excess salts and high biomass production in this condition. In addition, they must accumulate high levels of salts in the aerial part, so that the salts can be removed when the plants are harvested. Halophytes are plants adapted to high levels of salinity in the soil and have the ability to accumulate high amounts of salts in their tissues (ZHU, 2001). Because of these characteristics, they can be used to recover soils affected by salts (SQUIRES & AYOUB, 1994; MIYAMOTO et al., 1996).

With regard to halophytes, the genus Atriplex has stood out in semi-arid regions due to its easy establishment, important protein support, constant forage production and good acceptability by cattle (AGANGA et al., 2003; BEN SALEM et al., 2003). A. nummularia can be used as an important forage resource to supplement ruminant diets, thanks to its nutritional value (around 17 % protein) and high digestibility (70 %) (IPA, 2004), and it also has soluble salt absorption rates of up to 1.15 t ha-1 year-1 (PORTO et al., 2001).

2.6.1 Salt grass (Atriplex numularia L.)

Atriplex nummularia L. (figures 1 and 2), commonly known as salt grass, originated in the desert regions of Australia and is considered a halophytic plant. This plant is one of the most important species used in the recovery of saline soils worldwide, contributing to the process of phytoremediation of these soils. Thus, phytoextraction of soluble salts using halophytic plants is a low-cost alternative for recovering saline soils (LEAL et al., 2008; FREIRE et al, 2009; LIMA JUNIOR et al., 2010).

Figures 1 Salt grass (Atriplex numularia L.) in the experiment area (Source: author).

The genus Atriplex belongs to the Chenopodiaceae family, which has more than 400 species distributed in the various arid and semi-arid regions of the world, mainly in Australia, where it has a diversity of species and subspecies (FRANCLET & Le HOUÉROU, 1971; FREIRE et al., 2010). It is a long-lived shrub that is one of the forage resources adapted to dry land in arid and semi-arid regions on several continents (FREIRE et al., 2010).

14

Atriplex nummularia is a plant that requires high concentrations of NaCl for its growth. It stands out in semi-arid regions due to its easy establishment, important protein support, constant production of fodder and good acceptability by cattle (AGANGA et al., 2003; BEN SALEM et al., 2003).

Saltgrass has a C4 metabolism, which is when the initial product of primary CO2 fixation from photosynthesis results in molecules with four carbons, while in C3 plants, compounds with three carbons originate initially. Under high temperature and light conditions, C4 plants are more efficient than C3 plants due to their physiological characteristics, which favour photosynthetic and transpiration rates, giving them high water use efficiency (TEIXEIRA et al., 1983).

Increased salinity results in greater succulence, conserving water inside the plant tissue and reducing the rate of transpiration. Among halophytic plants, Atriplex nummularia is the most salt-tolerant (MIYAMOTO et al.,1994).

Atriplex species are generally characterised by their ability to withstand both aridity and salinity and produce fodder rich in protein and carotene. They also have the property of maintaining abundant active leaf phytomass during unfavourable periods of the year, especially during the dry season (OLIVARES, 1983). Their use as fodder is important in supplementing ruminant diets, thanks to their nutritional value (around 17 per cent of

protein) and high digestibility (70%) (IPA, 2014), as well as soluble salt absorption rates that reach approximately 1.15 t ha^{-1} year^{-1} (PORTO et al., 2001).

2.6.2 Chickweed (Eleusine indica L.)

Chickweed (Figures 3 and 4) is a herbaceous plant from the grass family, known mainly as an invasive weed. It is a member of the Poaceae family and belongs to the weed group. It grows in continental, equatorial, Mediterranean, oceanic, subtropical, temperate and tropical climates and is of Asian origin, reaching heights of 0.4 to 0.6 metres.

Figure 2. Chicken-foot grass (Eleusine indica L.) in the experimental area (Source: author).

Hen's-foot grass grows well in any type of soil and is a prominent presence on roadsides and wastelands, as well as infesting various crops. Relatively resistant to drought and high humidity. It is mainly spread by animals, as its seeds are not digested. Today it is distributed throughout the world's tropical, subtropical and temperate regions. In Brazil, it is found throughout the country, being common in the south, southeast, centre-west and the firm lands of the Amazon region (IVES GOULART, 2013).

Eleusine indica L. has great plasticity, its seeds germinate at any time of the year, but in winter growth is slower although seed production is equally high. There are cases of resistance to certain herbicides in the centre-west for this species (IVES GOULART, 2013).

2.6.3 Parsley *(Ipomoea asarifolia* L.)

There are several species of the Ipomoea genus and they are among the weeds that frequently occur in cultivated areas (BLANCO, 1978; GROTH, 1991; KISSMAN & GROTH, 1992). Ipomoea asarifolia is a perennial liana with a creeping habit, from the Convolvulaceae family, popularly known as parsley, potato parsley, batatarana, salsa- brava, salsa da praia, which is found among various crops and on the banks of rivers, lagoons and sea beaches, on abandoned land, on roadsides and in places close to houses, preferably on sandy soils (BLANCO, 1978).

Figure 3 - Wild sage *(Ipomoea asarifolia* L.) in the experimental area (Source: author).

Parsley is a plant with a soft or malleable stem, usually creeping, without the presence of lignin, i.e. without a woody stem (JUNIOR et al., 2008). According to AUSTIN & CAVALCANTI (2008) and DOBEREINER et al. (1960), in terms of its distribution and habitat, this Convolvulaceae is pantropical, with a wide occurrence in Brazil, mainly in the North and Northeast regions, where it is widely used for dermatological treatment, has high fruiting and germination rates, a high number of seeds per fruit and vegetative propagation, parsley can be considered an "ideal invasive plant", according to the terminology described by Baker (1974).

Ipomoea asarifolia L. is abundant in the Maranhão wetlands, where it remains green all year round and acts, in some areas, as a natural barrier to the advance of the dunes (CHAVES et al.,

2008).

2.6.4 Agricultural Gypsum

Gypsum is a mineral with various hydration products, the most common of which is gypsum [Ca(SO4)2.2H2O]. Its efficiency is determined by the degree of dissolution, gypsum is an ore that occurs abundantly throughout the world and whose solubility is around 2.04 g L^{-1} at 25°C, which in turn is influenced by the method of application and the particle size composition of the gypsum particles (BARROS et al., 2004).

According to Dias (1992), the Ca^{2+} ion supplied by gypsum can react in the soil exchange complex, displacing Al^{3+}, K^+ and Mg^{2+} into solution, which can in turn react with sulphate ($SO4^{2-}$) to form $AlSO^{4+}$ and the neutral ionic pairs $K2SO4^0$ and $MgSO4^0$ in addition to $CaSO4^0$, which are highly mobile in the profile. In this way, gypsum promotes greater movement of cations in depth due to factors such as its solubility, maintenance of the electrical charges of the soil subjected to gypsum plastering, and the fact that a large part of the sulphate anion remains in the soil solution (ERNANI, 1986; CAIRES et al., 1998).

In view of the information mentioned above, it can be seen that gypsum does not correct the soil's pH, given the presence of the $SO4^{2-}$ anion in the solution. Thus, unlike the carbonate anion from limestone, $SO4^{2-}$ does not receive protons (H^+), which means that the pH of the soil does not change (RAIJ, 2008). Even without any impact on pH, the benefits of plastering should be considered, since the input promotes improvements in the chemical and physical characteristics of the soil, thus improving the mineral nutrition of plant species and increasing the productivity of commercial crops (SOUSA et al., 1996; CAIRES et al., 2002 and SORATO et al., 2010).

There are several correctives that can be used to recover soils with excess exchangeable sodium, and gypsum is the most widely used corrective due to its simplicity of handling, ease of finding on the market and low cost (BARROS et al., 2009). The main areas where gypsum occurs in Brazil are the reserves associated with the sedimentary basins, especially the Araripe basin (MELO et al., 2008). The state of Pernambuco has a production of 2.6 million tonnes year-1 in the Pólo Gesseiro region, representing 95% of all Brazilian mineral gypsum (FREIRE et al., 2007).

In view of the recognised efficiency of gypsum and knowing that salinisation or sodification of the soil is responsible for a reduction in agricultural production, almost always culminating in the abandonment of arable areas, which causes damage to the regional economy (MELO et al., 2008), gypsum is the most widely used corrective in the recovery of sodic and saline-sodic soils. When designing a soil reclamation project using this chemical corrective or other techniques, it is important that factors related to the characteristics of the region and the soil, as well as the conditioners to be used, are assessed and compatible.

In soils with excess salt problems, gypsum reduces surface sealing, helping to increase the rate of water infiltration into the soil and consequently reducing Na levels$^+$ in the exchange complex. The criteria established in the process of remediation and management of soils affected by salts are based on the movement of water in the soil (CAVALCANTE et al., 2010), and aim to provide favourable conditions of humidity, aeration and salt balance for the root system of crops.

It is therefore necessary to install an underground drainage system in irrigated areas in order to leach the salts, which is one of the main infrastructures in the soil desalination process, due to the low permeability of these soils (ARAÚJO et al., 2011).

Therefore, gypsum is a conditioner that is widely used in the leaching of soils with salt problems, as well as being low-cost and abundant, but its leaching efficiency is greater the better the drainage system.

In the chemical correction treatment with gypsum application, the need to apply the gypsum corrective was based on the Theoretical Dose equation for the need for gypsum (Table 4). The amount of gypsum was determined based on the initial soil analysis, taken before the experiment was set up, which aimed to lower the PST from 17% to 7%. All the gypsum was applied by hand, using a hoe, to a depth of 15 cm. The control treatment was left unmanaged and used as a benchmark against which to compare the other treatments.

Table 4. Gypsum requirements (PIZARRO, 1978)

$$NG \ (kg/ha) = ((PST_1 - PST_2) \times CTC \times 86 \times pr \times ds) / 100$$

$PST_1 = 17$: value found in the calculation, in relation to the sodium content in the soil analysis; $PST_2 = 7$: value we want to achieve to reduce the sodium content in the soil; soil CTC= 8.2; CTC = Ca + Mg + K + Na + (H + Al); 86 = equivalent weight of agricultural gypsum CaSO4.2H2O = 40+32+(16x4) + 2 x (1x2 + 16) = 172 divided by valence 2 = 172/2 = 86; pr = 0.15: depth to be corrected in cm; ds = 1.53: soil density.

CHAPTER 3

MATERIAL AND METHODS

The work was carried out in the Irrigated Perimeter of São Gonçalo - PB, and before setting up the experiment, chemical (fertility), salinity and physical analyses of the soil were carried out in order to diagnose the quality of the soil and then use techniques and methods aimed at recovering the area degraded by sodification using phytoremediation techniques with Atriplex numulária L., Eleusine indica L., Ipomoea asarifolia L. and Gypsum.

agricultural. The data was analysed by variance and the means were subjected to the Tukey test at 5% probability.

The operations involved in setting up the experiment included: isolating the area, manually clearing the area, turning the soil by hand to incorporate the gypsum and planting the seedlings of Atriplex nummuralia L., Eleusine indica L. and Ipomoea asarifolia L. The control treatment was kept without any management, serving as a benchmark against the others. The seedlings of the species were planted at a spacing of 0.50 x 0.50 m, totalling nine plants per plot.

The need for gypsum was estimated according to the clay content of the soil, the exchangeable sodium content and the depth of the soil to be corrected, according to Richards' methodology (1954), at a depth of 0-15 cm. The gypsum was applied by hand and incorporated into the soil forty days before the seedlings were transplanted, during which time the soil was kept moist for the desired chemical reaction of the gypsum. Before this, however, trenches were dug with the appropriate slope to install the drains. At the end of the 40-day period, two sheets of water were applied to all the plots to leach out the salts, after which the plants were transplanted. The plants were irrigated with water from the Amazon well, the physico-chemical analysis of which was carried out before the start of the experiment, applying a rate of 8 mm at each irrigation three times a week.

The plants were pruned at a height of 15 cm from the ground every 70 days, in three collections over a period of 210 days, at which time the production of green and dry phytomass of the aerial part was determined. At the same time, soil samples were collected from the plots for analysis of pH, EC, Na, Ca, Mg, calculation of PST, calculation of RAS, apparent density, total porosity and degree of flocculation.

3.1 Diagnosis and soil quality of the São Gonçalo Irrigated Perimeter - PB

3.1.1 Characterisation of the Study Area

The study area is located in the Irrigated Perimeter of São Gonçalo - PB (Figure 7), situated at an altitude of 220 metres, between the geographical coordinates Latitude: 6° 46' 4" S and Longitude: 38° 12' 36" W. The average annual rainfall is around 894 mm, with the rainy season extending from January to May MASCARENHAS et al., 2005).

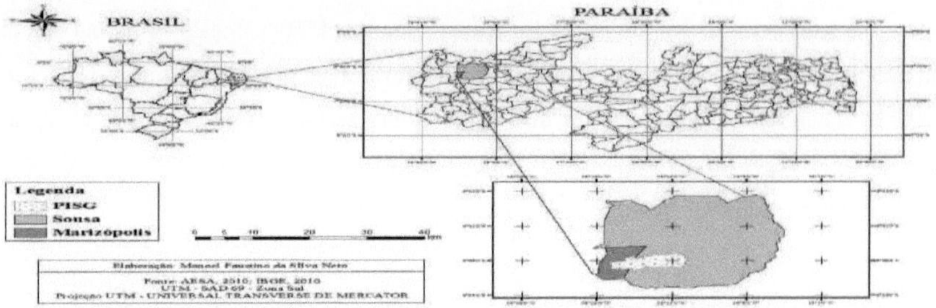

Figure 4 - Geographical location of the experimental area in the São Gonçalo-PB Irrigated Perimeter

São Gonçalo's climate is characterised as hot semi-arid (type Bsh in the Koppen climate classification), characterised by the scarcity and irregularity of rainfall, as well as strong evaporation due to the high temperatures. Average temperatures vary, with lows of 22 °C and highs of 38 °C. The rainy season lasts only five months, with average annual rainfall of almost a thousand millimetres, while annual evaporation is around 1500 millimetres (figure 8).

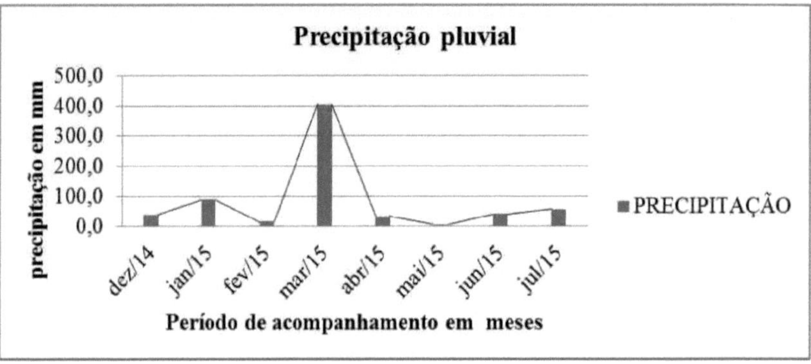

Figura 5. Rainfall from December 2014 to July 2015 at the experiment site (Source: author).

During the monitoring period, irrigation was carried out to make up for the lack of or insufficient or poorly distributed atmospheric rainfall. In this period, therefore, an 8 mm layer was maintained every two days, suspending it on days when rainfall exceeded this level. Temperatures were monitored daily during the experiment (Figure 6). The maximum and minimum temperatures were 45°C and 15°C respectively at the research site.

Figura 6. Average, maximum and minimum temperature values during the period of the experiment, measured daily. Note: The measurements were taken at 9am at the time of irrigation. (Source: author)

3.2. TREATMENTS AND DESIGN

The experiment was conducted in a randomised block design, consisting of eight treatments and four replications, the treatments being: soil cultivated with Atriplex nummularia L. plants; soil cultivated with Eleusine indica L. plants; soil cultivated with Ipomoea asarifolia L. plants; soil without cultivation with agricultural gypsum application; agricultural gypsum application + cultivation of Eleusine indica; agricultural gypsum application + cultivation of Ipomoea asarifolia L.; soil without cultivation with the application of agricultural gypsum; application of agricultural gypsum + cultivation of Atriplex; application of agricultural gypsum + cultivation of Eleusine indica; application of agricultural gypsum + cultivation of Ipomoea asarifolia L. and soil without management (control). The area of the experiment was flat, all the experimental plots were 1.96 m^2 (1.40 x 1.40 m), with a total area of 62 m^2 . All the plants were planted at a spacing of 0.50 x 0.50 m, with nine plants per plot (Figures 7A and B, 8 A and B, 9 A and B). Source: the author.

Figures 7 A and B. (Atriplex nummuralia L.) (Source: author)

Figures 8 A and B. (Eleusine indica L.) (Source: author)

Figures 9 A and B. (*Ipomoea asarifolia* L.) (Source: author)

3.3. DESCRIPTIVE STATISTICS AND ANALYSIS OF VARIANCE

The variables evaluated were analysed for variance according to the factors treatment and time, considering those that were significant. The means of the significant variables were subjected to the Tukey test (P < 0.05), using SISVAR software (FERREIRA, 2011).

3.4. SETTING UP AND CONDUCTING THE EXPERIMENT

The experiment was conducted in the Irrigated Perimeter of São Gonçalo-PB, in the area of the Federal Institute of Education, Science and Technology of Paraíba from 14 December 2014 to 14 July 2015. The soil in the experimental area was classified as a saline-sodic Eutrophic Chromic Luvissolo (EMBRAPA, 2013; RICHARDS, 1954) and collected at a depth of 0-15 cm. The soil was air-dried, crumbled, homogenised and passed through a 2 mm sieve for physical and chemical characterisation after the TFSA (Fine Air-Dried Soil) preparation procedure (Figure 10 A and B).

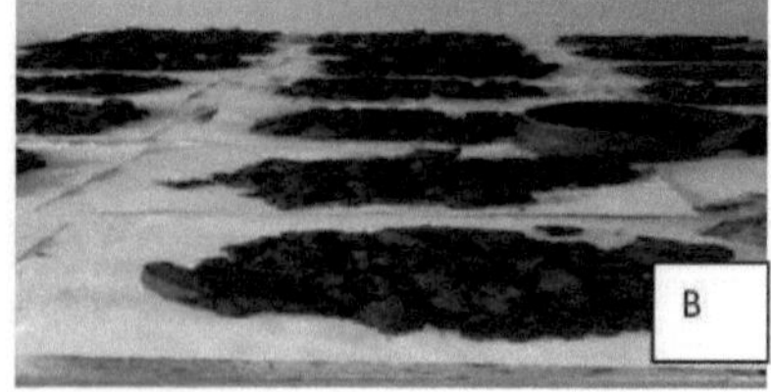

Figures 10 A and B. Drying the soil samples to prepare the TFSA. (Source: author)

Four samples were taken, the first before the experiment was set up and at 70, 140 and 210 days after the treatments were applied. Soil samples were taken from the 0-15 cm depth to carry out physical and salinity analyses. Four composite samples were taken from each plot and a single sample was taken for laboratory analysis in order to diagnose the soil's chemical and physical attributes (Figure 11 A and B).

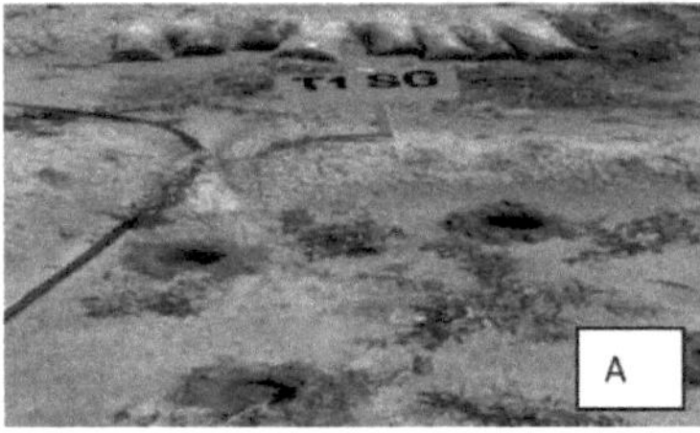

Figures 11 A and B. Soil sampling after pruning (Source: author)

3.4.1 Evaluation of soil quality in the São Gonçalo Irrigated Perimeter - PB

Soil characterisation was carried out on composite soil samples before the experiment was set up, then single samples were taken, prepared and analysed, and the saturation extract was obtained according to Richards' methodology (1954). Electrical conductivity (EC) was determined in the extract of the saturated paste, pH in the paste, calcium and magnesium by titration and sodium and potassium by flame photometry. The values obtained for calcium, magnesium and sodium were used to quantify the Sodium Adsorption Ratio (SAR) and to find the estimated PST. Before the experiment was set up, the soil was analysed in order to diagnose and classify the soil, as the pH in the saturated paste was equal to 8.5, the electrical conductivity in the saturation extract was greater than 4.0 dSm^{-1} and the PST was greater than 15%, therefore it was classified as Saline-Sodic, according to Richards, 1954. The results are shown below (Table 5).

Table 5. Soil analysis and classification in the experiment area at PISG

				Soil Salinity Analysis								
Depth cm	% of water in the ps	pHps	CEes dSm^{-1}	K$^+$	Na+	Ca+2	Mg+2	CO3^{-2}	HCO-3	Cl	PST	*Soil Class
						mmolcL ----						
0-15	28,5	8,5	4,23	0,42	24,2	4,6	1,45	0,00	39,00	35,0	17	Saline-Sodium

*Richards, 1954

PISG: Perímetro Irrigado de São Gonçalo - PB; pH$_{ps}$: pH in the saturated paste; EC$_{es}$: electrical conductivity in the saturation stratum; PST: Percentage of soluble sodium in the sorbent complex.

In the physical characterisation, before the research was set up and following EMBRAPA (1997) procedures, the granulometric composition, particle density by the volumetric ring method and particle density by titration in ethanol were determined (Table 6).

Table 6. Physical analysis and textural class of the soil in the experiment area at PISG

Physical Soil Analysis

Profesor cm	Particle size -------- gkg^{-1} ----------		Dens. Apparent g cm^3	Dens. Actual g cm^3	EN m m^{33}	Humidity Mpa 0,01 gKg-1	0.033	1,5	Available water gKg^{-1}	Natural clay gKg^{-1}	GF gKg^{-1}	Class Textural
0-15	720 205	75	1,53	2,84	0,86	155	181	70	85	50	333	Sand-France

PISG: São Gonçalo Irrigated Perimeter - PB; Granulometry: Clay and Silt by Boyouccos densimeter, Sand by sieving; Relative density: volumetric ring method; Real density: ethanol flask method; Moisture: Estimate based on textural class; PT: total porosity; GF: degree of flocculation.

3.4.2. Irrigation water analysis and quality

Prior to setting up the experiment in the field, physical and chemical analyses of the irrigation water were carried out during the dry season (December 2014) and soon after the rains, drainage water was also analysed (Table 7).

Table 7. Physico-chemical analysis of irrigation water

Source	PH	EC dSm^{-1}	K+	Na+	Ca+2	Mg+2	NaCl	CaCO3	RAS	Class
			-------------- mmol L$_c$ -----------				mgL-1 (mmol$_c$ L)^{0,5}			
Amazonas well (dry season)	7,9	0,39	0,12	0,68	0,85	0,90	179,1	153,8	0,73	C2S1
Amazonas well (after the rains)	8,6	0,50	0,17	1,02	1,30	0,90	225,0	196,6	0,97	C2S1

C2 S1: Water of medium salinity and low sodium content, suitable for irrigation in soils with good drainage, with no risk of causing salinisation or alkalinisation, although sodium-sensitive plants can accumulate harmful amounts of sodium; EC: electrical conductivity; SRA: sodium adsorption ratio.

In the specific case of the Northeast, the water used for irrigation comes from rivers, reservoirs and tube wells and, with a few exceptions, has EC values below 0.75 dS m^{-1} and a sodium percentage below 60 per cent. These waters are considered to be of good quality and do not present major problems for irrigation under suitable management conditions. However, as a result of the inadequate salt balance, commonly due to a lack of drainage, there is a gradual salinisation of the irrigated soil profile and a progressive increase in problem areas (EMBRAPA SEMIÁRIDO, 2001).

When assessing the quality of irrigation water in relation to sodium danger, the Sodium Adsorption Ratio (SAR) and electrical conductivity are taken into account.

CHAPTER 4

CONDUCTING RESEARCH IN THE FIELD

4.1. Leaching Blades

Before transplanting, the gypsum was sprayed on the treatments with gypsum and all the plots were kept moist to allow the chemical reaction of the gypsum to take place. The same conditions of humidity were applied to the other plots for 40 days. At the end of this period, two leaching sheets (Figure 12) were applied to leach out the salts. A small substation was installed in the experimental area with a maximum and minimum thermometer and a rain gauge to monitor the temperature and rainfall on a daily basis (Figure 13).

Figure 12. Leaching slides (Source: author) **Figure 13**. Rain gauge and thermometer (Source: author)

4.2. Plant treatments

In the phytoremediation treatments, seedlings of Atriplex nummularia L., Eleusine indica L. and Ipomoea asarifolia L. aged 60 days were used. The Atriplex nummularia L. and Ipomoea asarifolia L. seedlings were propagated by cuttings and the Eleusine indica L. seedlings by rooting with tillers from a mother plant. All the seedlings were propagated in a substrate prepared from organic compost and sand (1:3). Transplanting was carried out manually, with one plant per hole, placing only the seedling's root ball, without applying fertilisers at any time. Immediately after transplanting the seedlings, manual irrigation was carried out in the hole to stabilise the seedlings, maintaining a daily irrigation of 4 mm for the first 30 days to establish the seedlings.

Weeds were controlled by weeding with a hoe. There was a caterpillar attack on Atriplex nummularia L and Ipomoea asarifolia L after the first harvest, which was controlled by spraying with an insecticide based on Methomyl.

4.3. Laboratory analyses

4.3.1. Soil Chemical Attributes.

The soil in the experimental area was analysed every 70 days, in a total of three samples, at depths of 0-15 cm, after obtaining the saturated paste extract to determine pH, electrical conductivity (EC at 25°C), the content of soluble cations calcium, magnesium and sodium. The results of the analyses will be used to calculate the sodium adsorption ratio (SAR) and the percentage of exchangeable sodium (PST). To assess the chemical attributes, the soil samples were analysed for soluble elements by preparing the saturated paste extract using the method described by EMBRAPA (1997). Electrical conductivity (EC at 25°C) and pH were measured in the saturated paste extract, and soluble cations were determined: Ca^{2+} and Mg^{2+} , by titration; Na^+ and K^+ by flame emission photometry.

4.3.2. Preparation of the Saturated Paste

After the soil had been pulverised and the TFSA prepared, the saturated paste was prepared (Figure 13). The pH was measured in the paste, the ECes was measured in the saturation layer using a digital conductivity meter, the Ca^{2+} , Mg^{2+} contents were found by titration, Na^+ and K^+ by flame photometer.

Figure 13. Preparation of Saturated Paste (Source: author)

For chemical characterisation, the saturated paste was obtained according to the methodology described by Richard (1954), determining the electrical conductivity in the saturated paste extract (CEes) and the soluble ions of calcium and magnesium by titration, and sodium and potassium by flame photometry. After obtaining the soluble calcium, magnesium and sodium values, the sodium adsorption ratio (SAR) was quantified using the equation shown below. (Figure 14). We also analysed the exchangeable cations Ca^{2+} and Mg^{2+}, as well as K^+ and Na^+, extracted with a KCl 1 mol L solution^{-1} and Mehlich 1, respectively, and calculated: the sodium adsorption ratio (SAR), (Figure 14) and the percentage of exchangeable sodium (PST), (Figure 15), according to EMBRAPA (1997).

$$RAS = \frac{Na^+}{\left(\dfrac{Ca^{2+} + Mg^{2+}}{2}\right)^{1/2}}$$

Figure 14. Represents the equation for the Sodium Adsorption Ratio

Once the RAS was available, PST was calculated as a function of RAS based on the equation shown below. (Figure 15).

$$PST = \frac{100(0,01475RAS - 0,0126)}{1 + (0,01475RAS - 0,0126)}$$

Figure 15. Represents the equation for PST as a function of RAS

4.3.3. Soil Physical Attributes

Total clay and natural clay content were analysed using the densimeter method according to EMBRAPA (1997). Apparent density was analysed using the undeformed sample method and total porosity using water saturation.

4.3.4. Aspects of soil physical analyses

4.3.4.1. Particle size analysis

The granulometric analysis was carried out using the densimeter method according to EMBRAPA (1997), but with the use of a metal cup stirred with an electric stirrer for 5 minutes. Stir the suspension for 20 seconds with a stick, the lower end of which has a rubber cap with a slightly smaller diameter than the cylinder or beaker. Mark the time after the stirring has finished, take the reading with a densimeter after 2 hours, then remove the sand by washing the water through a 0.2 mm mesh sieve, fine sand and dispersed clay and silt are found by difference.

4.3.4.2. Clay dispersed in water

Water-dispersed clay (WDA) was determined following the method described for granulometric analysis, with stirring and reading after 2 hours and eliminating the use of chemical dispersants (EMBRAPA, 1997). Using the ADA and total clay data, the degree of dispersion was

27

calculated.

4.3.4.3. Density of the soil and its solid particles

Samples collected in rings (volumetric ring method) from each soil layer studied were used to determine soil density. The density of the solid soil particles was assessed using the volumetric flask method, using alcohol as the penetrating liquid. Both procedures were carried out according to EMBRAPA (1997).

The plant seedlings were prepared prior to installation in the field. The seedlings of Atriplex numularia and Parsley (Ipomoea asarifolia) were made by cuttings, and the grass (Eleusine indica) by rooting (Figure 14). After preparing the soil, the seedlings were transferred to the field and planted at a spacing of 50 cm between rows and 50 cm between plants.

The first 30 days were spent manually irrigating the hole to stabilise the seedlings, after which the micro-sprinkler irrigation system was installed, maintaining two irrigations a week. No fertilisation was carried out in the field, maintaining the initial fertility of the soil.

Caterpillars were sprayed to combat them. Every 70 days, the soil and plants were evaluated. The soil was analysed for physical properties and salinity, and the plants were measured for green and dry phytomass.

Figure 16 A and B. Preparation of seedlings of Atriplex numulária L., Ipomoea asarifolia L., Eleusine indica L. (Source: author)

CHAPTER 5

RESULTS AND DISCUSSIONS

5.1. Chemical attributes of the Irrigated Perimeter soils

In characterising the soil in the Irrigated Perimeter area, we can see that the soil initially had a high pH, a very high sodium content, PST in the range of soil with salinisation problems, but after applying the treatments there was an improvement in the soil's chemical and physical characteristics (Table 8).

It was found that the average ECes of the soils over the three sampling periods was mostly low, with the highest average value of 1.60 dS m^{-1} . However, the variability of the ECes was observed in all the samples analysed, with minimum and maximum values varying between the treatments applied, ranging from 1.16 to 2.33 dS m^{-1} , with an increase of more than 2 times the minimum value, so it can be concluded that despite this doubling of the variability in terms of ECes, most of the samples have values of less than 2.0 dS m^{-1} , being below the limit of 4 dS m^{-1} , which classifies soils as saline (USSL Staff, 1954). This reduction can be considered significant, as the lowest ECS value is twice as low as the highest ECS (2.33 dS m^{-1}), which is the untreated soil (the control). This indicates that the other treatments reduced the ECS during this research monitoring period and considering the initial ECS value (4.23 dS m^{-1}) (Table 8), there was a large reduction after the treatments were applied.

The salinity and physical analyses of the soil at a depth of 0-15 are summarised after 210 years of evaluation, in three 70-day samples, where the chemical and physical parameters analysed are characterised by means of averages (Table 8).

Table 8. Average soil characterisation (n°= 96 samples) during the evaluation time in the experiment at 70, 140 and 210 days.

				Salinity Analysis and Soil Physics								
Days	**Profes sor**	pHps	**CEes**	**K**$^+$	Na+	Ca+2	Mg+2	**RAS**	**PST**	**GF**	**EN**	**Ds**
	cm		**dSm**$^{-1}$									
0	0-15	8,5	4,23	0,42	24,2	4,60	1,45	13,91	17,0	333	0,86	1,53
70	0-15	8,7	1,76	0,40	4,90	3,74	3,61	2,85	4,24	337	0,46	1,45
140	0-15	8,1	1,65	0,41	4,68	4,66	3,30	2,62	3,90	350	0,50	1,37
210	0-15	7,9	1,41	0,61	4,60	5,81	4,18	2,14	3,19	363	0,49	140

pHps: pH in the saturated paste; ECes: electrical conductivity in the saturation stratum; RAS: sodium adsorption ratio; PST: Percentage of soluble sodium in the sorbent complex; GF: degree of flocculation; PT: total porosity; Ds: soil density.

5.1.1. Descriptive statistics for soil chemical attributes

The results of the descriptive statistical analysis of the soil attributes in the study area, in the surface layer from 0 to 0.15 m and the calculation of RAS and PST, can be seen in the values for mean, standard deviation and coefficient of variation. The data obtained from the soil analysis

(Table 9) showed high initial values of electrical conductivity of the saturation extract (ECes) for all the treatments evaluated in the characterisation of the area.

This can be attributed to the lack of drainage in the area during this period. However, after the drainage system was built and after the rainy season, there was a downward increase in EC values, with reductions reaching non-significant points during the period evaluated.

These reductions are explained by the leaching of soluble salts after the construction of the underground drainage system. This phenomenon is similar to the results found by Amaral et al. (2007). The parameters treatments and seasons together were not significant. The treatment factor alone was significant for most of the treatments, meaning that the treatments applied were important in the recovery of this soil, contributing to the improvement of the soil's chemical attributes.

Table 9. Summary of the analysis of variance for the soil's chemical attributes.

FV	GL	pHps	CEes	K^+	Na^{+2}	Ca^{+2}	Mg^{+2}	RAS	PST
Block	3	35,19 **	10,08**	3,12*	16,93**	6,13**	$0,73^{ns}$	20,66**	20,68**
Treatment	7	2,74**	2,62**	2,25*	7,50**	$1,97^{ns}$	$0,54^{ns}$	5,98**	5,98**
Season	2	140,87**	$1,46^{ns}$	$2,28^{ns}$	$0,17^{ns}$	6,95**	$1,87^{ns}$	$2,61^{ns}$	$2,61^{ns}$
Treatment x Season	14	$0,90^{ns}$	$0,72^{ns}$	$0,26^{ns}$	$0,47^{ns}$	$0,77^{ns}$	$1,10^{ns}$	$0,42^{ns}$	$0,42^{ns}$
CV %		2,49	52,81	95,73	44,70	47,04	49,63	49,49	49,48

*: $p < 0.05$; **: $p < 0.01$; ns: not significant; pHps: pH in the saturated paste; ECes: Electrical conductivity in the saturation extract; K: Soluble potassium; Na: Soluble sodium; Ca: Soluble calcium; Mg: Soluble magnesium; RAS: Sodium adsorption ratio; PST: Percentage of soluble sodium in the sorbent complex.

In soils affected by salts, the pH reaches values close to neutrality or in the alkaline range, indicating the predominance of OH ions⁻ in the soil solution. The average values found for the soil pH of the saturated paste for almost all plots, when compared in the three seasons, were alkaline, increasing and reaching higher values in the soil with grass and gypsum 8.34 and the lowest value in the soil with gypsum without cultivation 8.08 (Table 10). In the other treatments, although the value was high, there was no significant difference. These results for soil pH, in the extract of the saturation paste on average, for most of the samples collected in the three seasons, were in the alkaline range, predominantly between values greater than 8.0, contrasting with the pH values commonly observed in Brazilian soils, in which case there was a reduction in soil pH compared to the initial value of the soil (pH = 8.5).

These results differ from those found by Chaves et al. (2004), who carried out a survey of the chemical properties of the soils of Ilha de Assunção, located on the São Francisco River in Cabrobó, PE. After collecting 1,053 soil samples at depths of 0-30 cm on 1,131 ha, these authors found a wide range of pH values, varying from 4.43 to 8.90, from strongly acidic to alkaline. However, the average values found on the island indicate that moderately acidic pH results (5.85) predominate in the area studied. While those found in the São Gonçalo Irrigated Perimeter are on average in the alkaline range (Table 10).

Table 10. Soil chemical attributes as a function of the treatments applied.

Treatment	pHps	CEes	K^+	Na^+	Ca^{2+}	$Mg2^+$	RAS	PST
Soil G	8,08 b	1.61 a b	0,66 a	4.73 bc	6,64 a	4,11 a	2,22 b	3.31 b
Solo SG	8.24 a b	2,33 a	0,64 a	8,26 a	3,74 b	3,71 a	4,42 a	6,58 a
Atriplex G	8.11 a b	1,17 b	0,31 a	3.78 bc	4.22 a b	3,74 a	2,01 b	3,00 b
Atriplex SG	8.33 a b	1.76 a b	0,76 a	5.87 ab	4.97 a b	3,91 a	3.17 ab	4.72 a b
Grass G	8,34 a	1,27 b	0,37 a	4.01 bc	4.85 a b	3,61 a	2,26 b	3,37 b
SG Grass	8.16 a b	1.86 a b	0,29 a	3,08 c	4.10 a b	3,54 a	1,67 b	2,49 b
Salsa G	8.14 a b	1.61 a b	0,36 a	4.36 bc	5.09 a b	2,86 a	2,44 b	3,64 b
Salsa SG	8.20 a b	1,16 b	0,29 a	3.53 bc	4.26 a b	4,04 a	1,97 b	2,94 b

Averages followed by equal letters do not differ by the Tukey test (p<0.05) pHps :pH in the saturated paste; ECes: Electrical conductivity in the saturation extract; K : Soluble potassium; Na: Soluble sodium; Ca: Soluble calcium; Mg: Soluble magnesium; RAS: Sodium adsorption ratio; PST: Percentage of soluble sodium in the sorbent complex.

The data obtained for pH, PST and EC after the application of the treatments indicate a reduction in these three variables with the levels of the initial saline concentrations in the soil. It is worth pointing out that according to Richards (1954), the limit value for separating saline and non-saline soil is 4.00 dS m^{-1} . However, many crops have their production reduced well before this value (Table 10).

In the three soil samples analysed, there was a reduction in RAS and PST values (Table 10). Most of the RAS results were below 4.0 (mmolc L)$^{-10,5}$ and PST below 7.0 %. The highest results came from the control (Soil without Gypsum, Soil SG),

for both RAS and PST, 4.42 (mmolc L)$^{-10,5}$ and 6.58 %, respectively, as expected. The other treatments hardly differed from each other according to the Tukey test (Table 10).

The values found for RAS were below the classification limit for sodic soils (RAS > 15.0), a value recommended by the USSL Staff (1954). In addition, they were well below the PST of 7%, which represents problem-free soil (RICHARDS, 1954).

Among the soluble cations evaluated, Na$^+$ had the highest concentrations, especially in the untreated soil (the control), reaching average values of soluble Na$^+$ of up to 8.26 mmolc L^{-1} . It was also observed that the lowest value (3.08 mmolc L^{-1}) was in the gypsum-free grass treatment (Capim SG), indicating that this grass is efficient at extracting Na$^+$, surpassing even Atriplex, a grass that extracts this element for its metabolism. Several researchers agree that these types of plants are characterised by high productivity, resistance to water deficit, greater efficiency in the use of light energy and need salt (sodium chloride) as essential elements for their metabolism. They grow especially during the spring and summer months and can reach heights of over three metres.

Depending on the species, it has an upright, branched from the base, columnar and herbaceous growth habit (FAO, 1996). Oliveira et al. (2002) assessed the morphological, physical, chemical and mineralogical characteristics of three soil profiles affected by salts in the Custódia Irrigated Perimeter - PE, whose agricultural use was interrupted due to soil salinity and sodicity problems. They found that among the soluble cations assessed, Na^+ was the one with the highest concentrations, finding values of 9.5 mmolc L^{-1} and 10.5 mmolc L^{-1} in the 0-15 cm layer in two of the profiles analysed. Most of the results for this element are below 15.0 mmolc L^{-1}. Even these concentrations, according to Ayers & Westcot (1999), are capable of causing severe damage to sensitive crops.

There was no difference in soluble K^+ according to the Tukey test. The lower levels of K^+ in the soil solution are the result of this balance between the other cations, related to the levels of Na^+, Ca^{2+} and Mg^{2+}, which predominate in the system. Santos et al. (2005) state that, in general, K^+ concentrations tend to be lower as the Na: Ca ratio in the soil solution increases.

This result was expected because gypsum is a source of calcium (Table 9). Most of these soluble cations are present in average values below 5.5 mmolc L^{-1}, however, in some of the soils analysed, maximum values of 37.00 mmolc L^-
^{1} and 40 mmolc L^{-1} and minimums of up to 0.24 mmolc L^{-1} and 0.10 mmolc L^{-1} for Ca^{2+} and Mg^{2+}, respectively. In this case, the lowest average was observed for Soil without Gypsum (gypsum-free control, Soil SG), with a Ca cation^{2+} of 3.74 mmolc L^{-1}, also an expected result, given that this plot of soil received no treatment. With regard to the Mg cation,$^{2+}$ did not differ according to the Tukey test.

When you look at the Ca^{2+} and Mg^{2+} values found, you can see that they are high and the soils would be able to meet the needs of most cultivated plants, contrasting with the values of these nutrients in Brazilian soils in general, which are characteristically highly weathered and poor in these elements. However, they are within the most common range for semi-arid soils, which often have high levels of these ions. However, it is worth highlighting the possible problems of absorption of these nutrients, given the higher levels of Na^+ in some of these soils, which may hinder the absorption of Ca^{2+} and Mg^{2+} due to competition in the environment, as well as problems of hydraulic conductivity in the soils promoted by the predominance of Na^+, a characteristically dispersive ion, especially in semi-arid soils with high levels of 2:1 clay minerals in the clay fraction.

The average Ca values^{2+} found in the soil plots of the São Gonçalo irrigated perimeter are above the minimum limit for the soil solution presented by Ayers & Westcot (1999), who comment that if the concentration of this cation in the soil solution is less than 2.0 mmolc L^{-1}, there is a high probability of a reduction in crop production, as well as a calcium deficiency in the plant.

In this respect, soils with high ratios of the Na^+ ion to Ca^{2+} and Mg^{2+} are highly susceptible to colloid dispersion processes, due to the saturation of the exchange complex with the former to the detriment of the latter two. Because Na^+ is monovalent and has a large hydrated radius, it promotes the dispersion of negatively charged clay minerals, promoting colloidal dispersion. As a result, the dispersed colloids are carried along the soil profile, clogging pores in the subsurface and forming denser, hardened horizons or layers that make it difficult for water and air to move deeper. Soils with this type of problem are very common in the São Gonçalo Irrigated Perimeter area and are easily degraded.

5.2. Physical attributes of the soils in the experimental area

5.2.1. Descriptive statistics of the soil's physical attributes as a function of treatment

In terms of the soil's physical attributes, it can be seen that there was greater stability of the soil aggregates with an increase in the degree of flocculation, this increase being seen in the blocks and treatments. This can be attributed to the plant cultivations, which may be due to the fact that the plant roots may have made it possible for the aggregate to have greater space, facilitating aeration and water infiltration in the soil (Table 11). Soil aggregation may also be accelerated by root exploration in the soil profile, which, in the process of growth, brings the particles closer together as the roots exert pressure on the mineral particles as they advance through the pore space. The absorption of water by the roots also causes drying in the region adjacent to the roots, promoting an increase in the cohesive force between the particles (ZONTA et al., 2006).

Table 11. Summary of the analysis of variance for the soil's physical attributes.

FV	GL	GF	EN	Ds
Block	3	$10,79**$	$0,36^{ns}$	$5,00**$
Treatment	7	$2,74**$	$0,74^{ns}$	$1,43^{ns}$
Season	2	$0,26^{ns}$	$35,86**$	$45,38**$
Treatment x Season	14	$0,77^{ns}$	$1,22^{ns}$	$1,36^{ns}$
CV%		40,15	3,50	2,46

*: $p < 0.05$; **: $p < 0.01$; ns: not significant; GL: Degree of flocculation; PT: Total porosity; Ds: Soil density.

Analysing the average composition of the Degree of Flocculation of the plots in the Irrigated Perimeter in question (Table 12), there was a predominant increase in the Degree of Flocculation of the treatments (Parsley with Gypsum and Parsley without Gypsum), compared to the other treatments (Table 10). This means that parsley with or without gypsum is an excellent plant for flocculating soil particles, i.e. improving soil aggregates. There were no significant differences in total porosity (PT) and soil density (Ds) (Table 12).

Table 12. Physical attributes of the soil as a function of the treatments applied.

Treatment	GF	EN	Ds
Soil G	339 a b	0,48 a	1,40 a
Solo SG	211 b	0,47 a	1,42 a
Atriplex G	326 a b	0,49 a	1,39 a
Atriplex SG	332 a b	0,47 a	1,41 a

	385 a b	0,48 a	1,40 a
Grass G	385 a b	0,48 a	1,40 a
SG Grass	370 a b	0,48 a	1,39 a
Salsa G	413 a	0,48 a	1,42 a
Salsa SG	426 a	0,48 a	1,39 a

Averages followed by equal letters do not differ according to the Tukey test (p<0.05); GF: Degree of flocculation; PT: Total porosity; Ds: Soil density.

Secondly, Prado and Natale (2003), studying the degree of flocculation of a red latosol, found that the highest GF occurred in native vegetation, followed by reforestation and the direct seeding system. However, in the surface layers (0 to 20 cm), reforestation showed a similar degree of flocculation to native vegetation. According to Metzner et al. (2003), in soil where there is no tillage, clay particles and organic matter participate as aggregants in flocculation.

From an agricultural point of view, Lemos & Silva (2005) state that flocculation is important for erosion control, as it favours the formation of stable aggregates or granules. This is justified by the greater permeability and penetration of water, which favours plant growth (BUCKMAN, 1979, LEMOS & SILVA, 2005).

On the other hand, gypsum plastering increased the amount of stable aggregates, with an increase in the degree of flocculation in the plots with gypsum, especially in the parsley treatments (Table 12). Agricultural gypsum is calcium sulphate dihydrate ($CaSO_4 . 2H_2O$). When in contact with the soil solution, it can displace Al, magnesium (Mg) and potassium (K) from the exchange complexes, releasing them into solution. It also has the ability to form $AlSO_4^+$, which is less toxic to plants (ZAMBROSI et al., 2007).

Ruiz et al. (2006), working with saline-sodic soils in Paraíba, found that the corrective used (gypsum) provided higher values for the infiltration rate than the samples that did not receive gypsum. The authors also mention that this result was dependent on the clogging of the pores by the dispersed clay in the control treatment. Similar results were found by Suhayda et al. (1997) when they observed an increase in the water infiltration rate and aggregation of soil particles after applying gypsum to saline-sodic soil in China; in the treatment with gypsum the infiltration rate was 8 cm min^{-1} while for the treatment that did not receive gypsum the infiltration rate did not exceed 1 cm min.$^{-1}$

5.3. Dry phytomass production in the experimental area

5.3.1. Descriptive statistics of dry phytomass as a function of treatments and seasons

The production of fresh phytomass was significant among the blocks, as well as considering the treatments and evaluation times, while the production of dry phytomass was significant.

the treatments (Table 13). This phytomass production can be attributed to the fact that these plants are adapted to soils with salt problems.

For the phytoextraction of salts to be successful in saline-sodic soils, the plants must show tolerance to excess salts and high biomass production in this condition. In addition, they must accumulate high levels of salts in the aerial part, so that the salts can be removed when the plants are harvested. Halophytes are plants adapted to high levels of salinity in the soil and have the ability to accumulate high amounts of salts in their tissues (ZHU, 2001). Because of these characteristics, they can be used to recover soils affected by salts (SQUIRES & AYOUB, 1994; MIYAMOTO et al., 1996).

Table 13. Summary of the analysis of variance for Fresh Biomass (FB) and Dry Biomass (DBM).

FV	GL	BF	BS
Block	3	5.51**	2,23[ns]
Treatment	5	2,23 [ns]	14,40**
Season	2	1,00 [ns]	0,39 [ns]
Treatment x Season	10	3,44**	0,61[ns]
CV %		15,27	19,67

**: $p < 0.01$; ns: not significant; BF: fresh biomass; BS: dry biomass.

Chickweed with gypsum (G-chickweed) and without gypsum (SG-chickweed) had the highest dry phytomass averages at 70 days after transplanting, which is explained by their rapid growth in the first 70 days, followed by parsley (Ipomoea asarifolia) with gypsum and without gypsum, (G-parsley) and (SG-parsley), respectively (Table 14). The application of gypsum did not influence the phytomass production of crow's foot grass and Atriplex with and without gypsum plastering, but the latter showed higher production with gypsum plastering at 140 days (Table 14). This increase in the phytomass of wild parsley (Ipomoea asarifolia) can be attributed to the capacity to store reserves in the root system, which leads to greater regrowth capacity and growth speed (Table 14). At 210 days there was no difference between the two.

The higher BS production of Salsa-Brava (Ipomoea asarifolia) and Capim-de-galinha (Eleusine indica) can possibly be attributed to the fact that they are better adapted to soils with salt problems and are more responsive to short-term plastering when compared to Atriplex. Because it produces more dry phytomass than Atriplex, without the need for plastering, and because it can be used for animal feed, chickweed stands out as the best option for areas affected by salinisation.

Saltgrass (Atriplex numularia), the benchmark for adaptation to salinised soil conditions, did not differ significantly from wild parsley (Ipomoea asarifolia) and chickweed (Eleusine indica), with or without plastering, respectively, during this evaluation period.

This reduction in phytomass at 140 and 210 days (Table 14) can be attributed to the reserve of the root system of these plants, and over time these plants may be losing their regenerative power after pruning, taking into account local soil conditions, high temperatures and high local salinity levels, so the plants should feel it and therefore lose some weight after pruning.

Table 14. Comparison of Dry Biomass (DM) production averages according to the treatments applied and collection times.

Treatment	BS at 70 days	BS at 140 days	BS at 210 days
Atriplex G	497.41 b	536.61 c	924.02 a
Atriplex SG	508.58 b	383.29 c	755.75 a
Grass G	1422.37 a	1998.32 a b	681.02 a
SG Grass	1414.84 a	1633.61 b	951.23 a
Salsa G	1058.89 a b	2495.16 a	949.70 a
Salsa SG	762.94 a b	1329.16 b	625.73 a

Averages followed by equal letters do not differ according to the Tukey test (p<0.05); BS: Dry Biomass.

The higher productivity in a short period of time observed in the Chickenfoot grass (Eleusine indica) and Parsley (Ipomoea asarifolia) at 70 and 140 days can be explained by the fact that this crop is adapted to soils degraded by salts during the period in which this work was carried out, surpassing Atriplex, a shrub with a longer life cycle compared to the other two plants and adapted to arid soils and soils with salt problems, which could be an alternative for long-term phytoremediation (Table 14). Saltgrass (Atriplex numularia), despite being a plant that is highly tolerable to salinity (BARROSO et al., 2006), showed an increasing rate of biomass production over time, in line with the values found for chickweed (Eleusine indica) and wild parsley (Ipomoea asarifolia).

The relative importance of the various growth maintenance processes has a wide range of responses, varying according to species or even between genotypes of the same species. They also differ in relation to the time of exposure to salinity, the salt concentration and possibly the local environmental conditions (ARAÚJO et al., 2006). Based on these results, Atriplex nummularia's ability to regrow can be seen, as observed by Souza et al. (2011), in their study with the aforementioned species, comparing pruned and unpruned plants and verifying the plants' significant ability to recover when subjected to pruning.

Munns (2005) states that the osmotic effect caused by the high salt concentration around the roots significantly reduces the growth rate of the aerial part of salinity-sensitive plants. A sudden increase in soil salinity causes the leaf cells to temporarily lose their turgour; there is also a decrease in the rate at which new leaves appear and lateral buds develop, reducing the number of lateral branches or shoots. These same authors also state that plants with the ability to tolerate osmotic pressure and the accumulation of Na^+ in their tissues show an increase in their ability to maintain growth. And they emphasise that the effects are expressed differently in young and old tissues. Greater osmotic tolerance will be manifested by an increase in the capacity to produce new leaves, while tissue tolerance is revealed mainly by an increase in the survival of older leaves.

Halophytes have a controversial physiology when it comes to the effects caused by the presence of salts in the environment. This group of plants has different strategies for maintaining the integrity of its development process (MUNNS & TESTER, 2008).

In the case of Atriplex nummularia, the species has abilities that ensure it can live in conditions of high salinity. By controlling the osmotic pressure, the plant balances the absorption of Na^+ with other ions in the cells, ensuring the movement of water in the plant and overcoming the low external water potential, thus safeguarding intracellular homeostasis (YOKOI et al., 2002; AZEVEDO et al., 2005, SILVA et al., 2009).

These reports are in line with research by Porto et al. (2006), who stated that Atriplex, like other halophyte species, has the ability to withstand not only high levels of salinity in the soil-water complex, but also to accumulate significant amounts of salts in its tissues. This fact is reinforced by Leal et al. (2008), who tested the phytoextraction potential of Atriplex and observed that the species has the capacity to grow in soil with an average EC of 25.94 dS m^{-1} and an average PST of 51.61%, demonstrating its potential to occupy areas degraded by salts where other plants would not be able to grow, and can be considered an economically viable alternative for recovering the productive capacity of these areas.

Therefore, the period assessed for Atriplex may have been too short for it to fully develop and become accustomed to saline environments, possibly because Atriplex has not reached its level of development and habitation, which is why it does not differ statistically. It can also be seen that Atriplex tends to increase its production over time, while parsley, and especially chickweed, tends to decrease, possibly because it has reached its full life cycle (Table 14).

The estimated dry phytomass production of Parsley (Ipomoea asarifolia) with gypsum (Parsley G) obtained the highest FS production with 2.92 Mg ha^{-1} (FS), followed by Chickweed (Eleusine indica) with gypsum (Chickweed G), with 2.64 Mg ha^{-1} , surpassing Atriplex. Chickweed (Eleusine indica) without gypsum (SG Grass) and parsley (Ipomoea asarifolia) without gypsum (SG Parsley), produced 2.55 Mg ha^{-1} and 1.75 Mg ha^{-1} , of FS, respectively. The lowest FS production was from Atriplex with gypsum with 1.26 Mg ha^{-1} and Atriplex without gypsum with 1.06 Mg ha^{-1} of FS compared to the other plants. It is known that the density of plants per area influences crop productivity. In the case of the Brazilian semi-arid region, there is no information on the optimum density for this crop when irrigated. According to the FAO (1996), a convenient density would be 1,600 plants per hectare, with an expected yield of between 1,000 and 1,500 kg of dry matter per hectare per year.

The literature shows variability between 2.9 and 10.0 Mg ha^{-1} in the dry matter yield of Atriplex numularia L., attributed mainly to the quality of the growing environment (O'LEARY, 1986). These differences are also influenced from the point of view of management, cultural practices and harvesting, such as spacing, water levels, cutting height and periodicity. In the case studied here, the dry phytomass production of Atriplex numularia L. was lower than that seen in the literature. This is due to the height and periodicity of the cut (three cuts every 70 days), which means that the plant may not have fully developed due to its longer life cycle compared to the other two plants studied. Generally, plants start to produce seeds between the 2nd and 4th year after planting, however, it is common to find plants that produce seeds as early as the 1st year (HERREIRA, el tal, 2000).

The higher yields seen in Eleusine indica and wild parsley can be explained by the fact that these plants are adapted to soils degraded by salts. During the period in which this work was carried out, statistically they did not differ from Atriplex, which is an adapted shrub with a longer life cycle compared to the other plants compared.

The dry biomass production of chickweed (Eleusine indica) reached 5.21 Mg ha^{-1} of dry biomass, in summer planting in the state of São Paulo (MURAISHI et al., 2005), while Franscisco et al. (2007), in Piracicaba (SP), sowing in September and managing at full bloom with early fertilisation, obtained a production of 6.375 Mg ha^{-1} of dry biomass. Under off-season conditions, in the south-west of Goiás, the yield of this crop was 9.48 Mg ha^{-1} (ASSIS et al., 2005) and in Piauí it was 2.89 Mg ha^{-1} of dry biomass (AZEVEDO & NASCIMENTO, 2002). The variations in dry biomass production of capim-pé-de-galinha in the different regions are due to the genetic material used, the planting season and the climatic conditions of each region. The dry phytomass production of crow's-foot grass (Eleusine indica) that we found was similar to that found in Piauí.

Checking the dry phytomass production of (Ipomoea asarifolia) with and without gypsum (Salsa G) of 2.92 Mg ha^{-1} and 2.55 Mg ha^{-1} well below the values found by Linhares et al. (2008), who studied jitirana (Ipomoea sericophylla) and found that it had an average phytomass production higher than the minimum established for adding a plant to a crop rotation system, which according to Darolt, (1998) is 6 Mg ha^{-1} . The maximum dry phytomass production of jitirana was 3.59 Mg ha^{-1} at 120 days of growth. The low dry phytomass production is due to the fact that jitirana has a low dry matter content, ranging from 5.82% DM at 15 days of age to 11.02% DM at the last phenological stage (120 days of age). These results were lower than those obtained by Ceretta et al (2002) when they assessed the dry matter production of three plants used as ground cover in winter, with nitrogen fertilisation, obtaining dry matter production of 7.36; 6.11; 5.16 Mg ha^{-1} , respectively in plants of black oats, black oats + vetch and turnip rape in 1996. The results obtained with (Ipomoea asarifolia) with gypsum were close to those found by Linhares et al. (2008). This is because parsley with gypsum is responsive to gypsum and adapted to soils with salt problems.

The results presented here are for the soil and edaphoclimatic conditions found in the São Gonçalo Irrigated Perimeter, and considering the evaluation period, which was 210 years in the field. There is therefore a need for further studies on different water and crop management to sustainably assess the potential of Atriplex, crowfoot grass (Eleusine indica) and parsley (Ipomoea asarifolia) for fodder production and salt removal in salinised areas.

CHAPTER 6

CONCLUSIONS

- Parsley *(Ipomoea asarifolia),* with or without plastering, was the plant that contributed most to increasing the degree of soil flocculation and, consequently, to greater stability of soil aggregates;
- Chickweed *(Eleusine indica),* as well as parsley *(Ipomoea asarifolia)*, can be an alternative for recovering soils affected by salts in a short space of time;
- *Atriplex numularia* can be a method of long-term soil recovery;
- The use of gypsum as a soil amendment proved favourable in improving chemical attributes;
- Parsley *(Ipomoea asarifolia)* is more responsive to the application of agricultural gypsum in terms of plant biomass production;
- In terms of biomass production in the short term, chickweed *(Eleusine indica)* and parsley *(Ipomoea asarifolia)* did not differ statistically from Atriplex, a reference plant for phytoremediation of soil affected by salts.

CHAPTER 7

ANNEX
Experimental design in the field

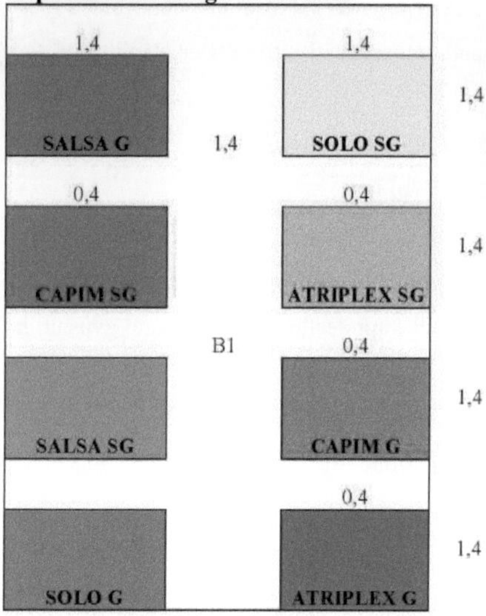

B1

1,4 — SALSA G 1,4 1,4 — SOLO SG
0,4 — CAPIM SG 0,4 — ATRIPLEX SG
SALSA SG 0,4 — CAPIM G
SOLO G 0,4 — ATRIPLEX G

Área total do Bloco: 23,04 m²

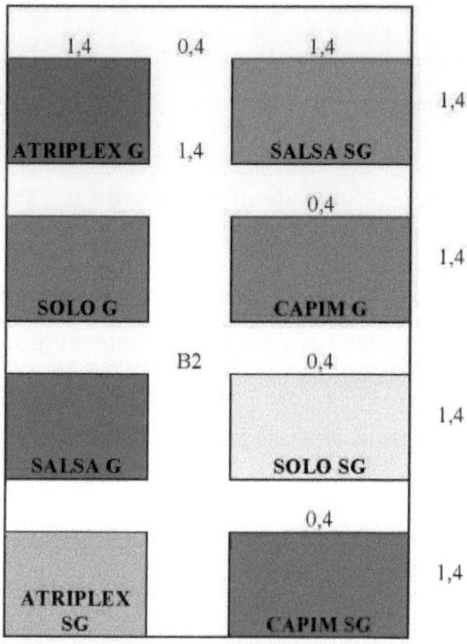

B2

1,4 — ATRIPLEX G 0,4 1,4 1,4 — SALSA SG 1,4
SOLO G 1,4 0,4 — CAPIM G 1,4
SALSA G 1,4 B2 0,4 — SOLO SG 1,4
ATRIPLEX SG 1,4 0,4 — CAPIM SG 1,4

Área total do Bloco: 23,04 m²

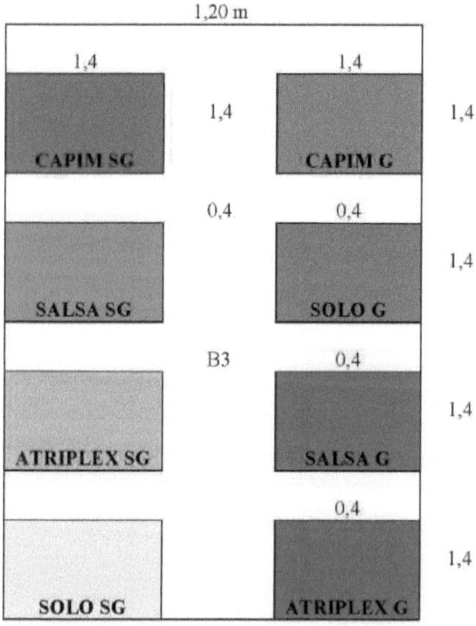

B3

1,20 m

1,4 — CAPIM SG 1,4 1,4 — CAPIM G
0,4 — SALSA SG 0,4 — SOLO G
B3 ATRIPLEX SG 0,4 — SALSA G
SOLO SG 0,4 — ATRIPLEX G

Total block area: 23.04 m²

B4

1,20 m

1,4 — SOLO SG 1,4 1,4 — CAPIM G 1,4
0,4 — ATRIPLEX G 1,4 0,4 — SALSA G 1,4
B4 SOLO G 1,4 0,4 — ATRIPLEX SG 1,4
SALSA SG 1,4 0,4 — CAPIM SG 1,4

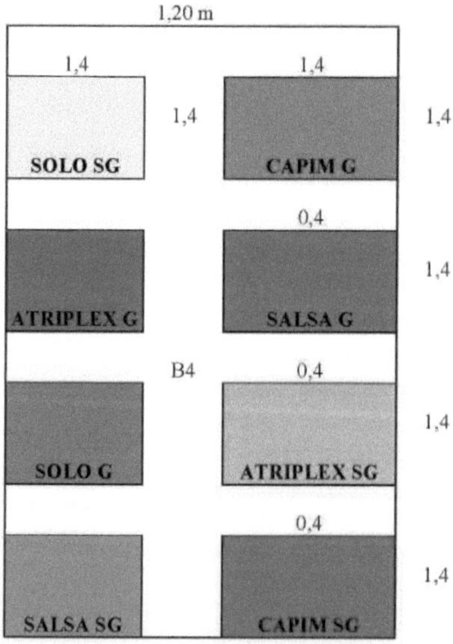

Total block area: 23.04 m²

CHAPTER 8

REFERENCES

AGANGA, A. A.; MTHETHO, J. K.; & TSHWENYANE, A. Atriplex nummularia (Old Man Saltbush): A potential forage crop for arid regions of Botswana. Pakistan Journal of Nutrition, v. 2, p. 72-75, 2003. Master's thesis.

AGUIAR NETTO, A. de O.; MACHADO, R. & VASCONCELOS, B. Diagnosis of the saline-sodification process in the Jabiberi-SE Irrigated Perimeter. Irriga, Botucatu, v. 11, n. 4, p. 448-459, 2006. Doctoral thesis.

AMARAL, F. C. S., Parahyba, R. B. V; Silva, F.H.B.B., Leite, A. P.; Batista, M. J.; Barros, J. C. Pedological characterisation and drainage studies of the Brígida, Caraíbas and Apolônio Sales irrigation perimeters, State of Pernambuco- Electronic data - Rio de Janeiro: Embrapa Solos, 2007, 68p.(Boletim de pesquisa e desenvolvimento/ Embrapa Solos; 116).

ARAÚJO, A. P. B. de; COSTA, R. N. T.; LACERDA, C. F. de; GHEYI, H. R. Economic analysis of the recovery process of a sodic soil in the Curu Irrigated Perimeter - Pentecoste, CE. Revista Brasileira de Engenharia Agrícola e Ambiental, v.15, n.4, p.377-382, 2011.

ARAÚJO, S. A. M. de; SILVEIRA, J. A. G.; ALMEIDA, T. D.; ROCHA, I. M. A.; MORAIS, D. L.; VIÉGAS, R. A. Salinity tolerance of halophyte Atriplex nummularia L. grown under increasing NaCl levels. Revista Brasileira de Engenharia Agrícola e Ambiental, v.10, n.4, p.848-854, 2006.

ASSIS, R.L.; MACEDO, R.S.; PIRES, F.R.; BRAZ, A.J.B.P.; SILVA, G.P.; PAIVA, F.C.; GOMES, G.V. & CARGNELUTTI FILHO, A. Dynamics of decomposition of species used as cover plants, cultivated in off-season, in the Cerrado of south-western Goiás. In: BRAZILIAN CONGRESS OF SOIL SCIENCE, 30, Recife, 2005. Proceedings. Recife, Brazilian Society of Soil Science, 2005. CD-ROM.

AUSTIN, D.F.; CAVALCANTI, P.B. Convolvulaceae of Amazonia. Publications BAKER, H. G. The evolution of weeds. Palo Alto, USA: Annual Review Ecology 2008. v. 1, p. 208-208.

AYERS, R.S. & WESTCOT, D.W. Water Quality in Agriculture. 2nd ed. Trad. GHEYI, H. R; MEDEIROS, J. F; DAMASCENO, F.A.V. Campina Grande. UFPB, 1999. 218p. FAO Studies: Irrigation and Drainage, 29.

AZEVEDO, C. M. da S. B.; PEDROSA, G. P.; MEDEIROS, J. F. de; NUNES, G.H. de S. Use of Atriplex nummularia in the extraction of salts from soils irrigated with saline effluents. Revista Brasileira de Engenharia Agrícola e Ambiental, v.9, p.300-304, 2005 (Supplement).

AZEVEDO, D.M.P. & NASCIMENTO, H.T.S. Potencial forrageiro de espécies para cultivo no período de safrinha em solos de tabuleiros costeiros. Teresina, Embrapa, 2002. 4p. (Technical Communication, 148).

BAKER, H. G. The evolution of weeds. Palo Alto, USA: Annual Review Ecology Systematics, v. 5, p. 1-24, 1974.

BARROS, M. de F. C.; BEBÉ, F. V; SANTOS, T. O. dos; CAMPOS, M. C. C. Influence of the application of gypsum to correct a saline-sodic soil cultivated with cowpea. Revista de Biologia e Ciências da Terra. v. 9, n. 1, p. 77 - 82, 2009.

BARROS, M. de F. C.; FONTES, M. P. F.; ALVAREZ, V. H.; RUIZ, H. A. Recovery of soils affected by salts by the application of quarry gypsum and limestone in the Northeast of Brazil. Revista Brasileira de Engenharia Agrícola e Ambiental, v.8, n.1, p.59-64, 2004.

BARROSO, D. D.; Araújo, G. G. L.; Porto, E. R.; Porto, F. R. Productivity and nutritive value of the forage fractions of saltgrass (Atriplex nummularia) irrigated with four different volumes of effluent from tilapia farming in brackish water. Revista Agropecuária Técnica, v.27, n.1, p.43-48, 2006.

BATISTA, M. J.; NOVAIS, F. de. & SANTOS, D. G. Drainage as a tool for desalination and prevention of soil salinisation. Brasília, SRH, 203p. 1998.

BEN SALEM, H.; NEFZAOUI, A. & BEN SALEM, L. Spineless cactus (Opuntia ficus indica f. inermis) and oldman saltbush (Atriplex nummularia L.) as alternative supplements for growing Barbarine lambs given straw-based diets. Small Ruminant Research, v. 51, p. 6573, 2003.

BERNARDO, S. Irrigation manual. 6ed. Viçosa; Federal University of Viçosa, 1995.

BERNARDO, S.; SOARES, A. A. & MANTOVANI, E. C. Manual de irrigação. 8 ed. Viçosa, MG: UFV, 2006. 625 p.

BLANCO, H. G. 1978. Catalogue of weed species infesting cultivated areas in Brazil. Bell family (Convolvulaceae). O Biológico 44: 259-278

BLANCO, H.G. Catalogue of weed species infesting cultivated areas in Brazil. Botânica, v. 41, p. 83-99, 1991. In Chaves, D. P. Experimental intoxication by Ipomoea asarifolia in sheep: clinical, laboratory and anatomopathological findings, Jaboticabal, 2009. Doctoral thesis.

CAIRES, E. F.; BARTH, G.; GARBUIO, F. J.; KUSMAN, M. T. Soil acidity correction, root growth and maize nutrition according to surface liming in a no-till system. R. Bras. Scielo. Solo, v.26, p.1011-1022, 2002.

CAIRES, E.F.; CHUEIRI, W.A.; MADRUGA, E.F. & FIGUEIREDO, A. Changes in soil chemical characteristics and soya bean response to surface-applied lime and gypsum in a no-tillage cropping system. R. Bras. Scielo. Solo, v. 22, p. 27-34, 1998.

CAVALCANTE, L. F; SANTOS, R. V. dos; FERREYRA F. F. H.; GHEYI, H. R. & DIAS, T. J. Recovery of soils affected by salts. In: GHEYI, H. R.; DIAS, N. S.; LACERDA, C. F. Manejo da salinidade na agricultura. Fortaleza, INCT Sal, 2010. p. 472.

CERETTA, C.A.; BASSO, C.J.; HERBES, M.G.; POLETTO, N.; SILVEIRA, M.J. Production and phytomass decomposition of winter soil cover plants and maize, under different nitrogen fertiliser management. Ciência Rural, Santa Maria, v.32, n.1, p.49-54, 2002.

CHAVES, D.P.; SOBRINHO, A.G.; MAHON, G.V.; CARVALHO, V.H.A; FAGLIARI, J.J. Outbreak of tremorgenic syndrome caused by Ipomoea asarifolia (Ders.) Roem. & Schult. Convolvulaceae) in sheep in Lençóis Maranhenses. In: XXXV BRAZILIAN CONGRESS OF VETERINARY MEDICINE, 2008, Gramado. Proceedings... Gramado-RS, Systematics, v. 5, p. 1-24, 1974. In Chaves, D. P. Experimental intoxication by Ipomoea asarifolia in sheep: clinical, laboratory and anatomopathological findings, Jaboticabal, 2009. Doctoral thesis.

CHAVES, L. H. G.; TITO, G. A.; CHAVES, I. B.; LUNA, J. G. & SILVA, P. C. M. Chemical properties of alluvial soil from the island of Assunção - Cabrobó (Pernambuco). Revista Brasileira de Ciência do Solo, v.28, p.431-437, 2004.

DAROLT, M.R. Principles for implementing and maintaining systems. In: No-till farming: sustainable smallholdings. Londrina: Iapar, 1998. p.16-45 (Circular, 101).

DIAS, N. S.; BLANCO, F. F. effects of salts on soil and plant. In: GHEYI, H. R.; DIAS, N. S.; LACERDA, C. F. In: Management of salinity in agriculture. Fortaleza, INCT Sal, 2010. p. 472. Extract from the Dissertation: Santos, Monaliza Alves dos, Recovery of saline-sodic soil by phytoremediation with Atriplex nummularia or application of gypsum. Recife, 2012.
do Museu Goeldi, v.36, p.1-134, 1982. In Chaves, D. P. Experimental intoxication by Ipomoea asarifolia in sheep: clinical, laboratory and anatomopathological findings, Jaboticabal, 2009. Doctoral thesis.

DOBEREINER, J., TOKARNIA, C. H., CANELA, C. F. C. Experimental intoxication by parsley Ipomoea asarifolia (R. et Schult) in ruminants. São Paulo: Arquivos do Instituto Biológico Animal, v.3, p.39-57, 1960. In Chaves, D. P. Experimental intoxication by Ipomoea asarifolia in sheep: clinical, laboratory and anatomopathological findings, Jaboticabal, 2009. Doctoral thesis.

EMBRAPA - Brazilian Agricultural Research Corporation. Manual of soil analysis methods. Rio de Janeiro, 1997. 41p.

EMBRAPA - Brazilian Agricultural Research Corporation. Embrapa Semi-Arido, Documentos 167. Petrolina - PE, 2001.

EPSTEIN, E.; BLOOM, A. J. Plant mineral nutrition: Principles and perspectives. Planta, ed. 2, 2006. 401p.

ERNANI, P.R. Changes in some chemical characteristics in the arable layer of the soil by the application of agricultural gypsum on the surface of native fields. R. Bras. Ci. Solo, v.10, p.241, 1986.

UNITED STATES. Department of Agriculture. Keys to soil taxonomy. 4.ed. Washington, USDA, 1990. 422p. (USDA. SMSS Technical Monograph, 6)

ESTEVES, B. S; SUZUKI, M. S. Effect of salinity on plants. Ecologia Brasileira, v.12, n. 4, p. 662-679, 2008.

FAGERIA, N. K. & GHEYI, H. R. Effects of salts on plants. In: GHEYI, H. R.; QUEIROZ, J. E.; MEDEIROS, J. F. de. (Ed.). Management and control of salinity in agriculture

irrigated. Campina Grande: UFPB/SBEA, P.125-131, 1997. Bell family (Concolvulacea). São Paulo: O Biólogo, v.44, p.259-278, 1978. In Chaves, D. P. Experimental intoxication by Ipomoea asarifolia in sheep: clinical, laboratory and anatomopathological findings, Jaboticabal, 2009. Doctoral thesis.

FAO. Global Network on integrated soil Management for Sustainnable Use of Salt-affected Soil. Rome: 2000. http://www.fao.org/ag/agl/agll/spush. Accessed 30 April 2014.

FAO (Rome, Italy). Case studies of plant species for arid and semi-arid areas in Chile and Mexico. Santiago: FAO Regional Office for Latin America and the Caribbean, 1996. 143p. il (FAO. Regional Office for Latin America and the Caribbean, Arid and Semi-arid Zones, 10).

FRANCLET, A. & LE HOUÉROU, H.N. The Atriplex in Tunisia and North Africa. Rome: FAO, 1971. 249 p. FAO Report Technique, 7.

FERREIRA, D. F. Sisvar - variance analysis system for balanced data. Lavras: UFLA, 2011.

FRANSCISCO, E.A.B.; CÂMARA, G.M.S. & SEGATELLI, C.R. Nutritional status and production of chickweed and soya grown in succession in an early fertiliser system. Bragantia, 66:259-266, 2007.

FREIRE, A. L. O.; RODRIGUES, T. J. D. Soil salinity and its effects on growth, nodulation and N, K and Na levels in Leucaena (Leucaena leucocephala (Lam.) de Vit.) Engenharia Ambiental, v. 6, n. 2, p. 163-173, 2009.

FREIRE, F. J.; FREIRE, M. B. G. S.; ROCHA, A. T. da, OLIVEIRA, A. C. de. Mineral gypsum from Araripe and its implications for agricultural productivity in sugarcane in the state of Pernambuco, Brazil. Anais da Academia Pernambucana de Ciência Agronômica, vol. 4, p.199-213, 2007. In: Santos, M. A. dos, Recovery of saline-sodic soil by phytoremediation with Atriplex nummularia or application of gypsum. Recife, 2012. 13 p. Master's dissertation.

FREIRE, M. B. G. S.; SOUZA, E. R. & FREIRE, F. J. Phytoremediation of soils affected by salts. In: GHEYI, H. R.; DIAS, N. S.; LACERDA, C. F. Manejo da salinidade na agricultura. Fortaleza, INCT Sal, 2010. p. 472.

GONÇALVES, I. V. C.; FREIRE, M. B. G. dos S.; SANTOS, M. A. dos; SOUZA, E. R. de; FREIRE, F. J. Chemical changes in a Fluvial Neossol irrigated with saline water. Revista Ciência Agronômica, v. 42, n. 3, p. 589-596, 2011.

GOULART, I. C. G. R.; Pé-de-Galinha-Eleusine *Indica* L., 2013: available at http://www.jardineiro.net/plantas/pe-de-galinha-eleusine-indica.html. Accessed in October 2015.

GROTH, D. Morphological characterisation of seeds and seedlings of seven weed species of Convolvulaceae occurring in agricultural seeds in Brazil. Porto Alegre: Iheringia, Série Botânica, v. 41, p. 83-99, 1991. In Chaves, D. P. Experimental poisoning by Ipomoea

asarifolia in sheep: clinical, laboratory and anatomopathological findings, Jaboticabal, 2009. Doctoral thesis.

HOLANDA, J. S. Management and use of salinised areas in the Açu valley. Bank of the Northeast. Fortaleza, 2000. 95p.

HORNEY, R.D.; TAYLOR, B.; MUNK, D.S.; ROBERTS, B.A.; LESCH, S.M. & PLANT, R.E. Development of practical site-specific management methods for reclaiming saltaffected soil. Comp. Electr. Agric. 46:379-397, 2005.

IPA - AGRONOMIC INSTITUTE OF PERNAMBUCO. Atriplex nummularia. Available at: <http://www.ipa.br/resp11.htm>. Accessed on 21 April 2014.

JUNIOR, D.A.O. de; SILVA, R.A. da; ARAÚJO, L.L.S. dos; JÚNIOR, R.J.S. dos; ARNAULD, A.F. Phenological characterisation of herbaceous and shrubby bee plants in the micro-region of Catolé do Rocha - PB - Brazil. Revista Verde de Agroecologia e Desenvolvimento Sustentável, v. 3, n. 4, p. 86-99, 2008. available: (http://revista.gvaa.com.br).

KEIFFER, C.H. & UNGAR, I.A. Germination and establishment of halophytes on brine- affected soils. J. Appl. Ecol., 39:402-415, 2002. Leal, I. G. et al., extracted from the article Phytoremediation of Sodic Saline Soil by Atriplex nummularia and Gypsum. Published in R. Bras. Scielo. Solo, 32:1065-1072, 2008.

KISSMAN, K.G.; GROTH, D. Weeds and noxious plants. São Paulo: Basf, 1992.

KOPPEN, W. Climatologia: con un estudio de los climas de la Tierra. Mexico, Fondo de Cultura Económica. 1948.

KOYRO, H. W. Effect of salinity on growth, photosynthesis, water relations and solute composition of the potential cash crop halophyte Plantago coronopus (L.). Environmental and Experimental Botany, v.56, n.2, p.136-146, 2006.

LEAL, I. G.; ACCIOLY, A. M. A.; NASCIMENTO, C. W. A.; FREIRE, M. B. G. S.; MONTENEGRO, A. A.; FERREIRA, F. L. Phytoremediation of saline sodic soil by Atriplex nummularia and gypsum from quarries. Brazilian Journal of Soil Science, 2008.

LEMOS, C.F. de; SILVA, E.T. da. Comparison of soil morphological, mineralogical, chemical and physical characteristics between no-till and conventional tillage areas. Revista Acadêmica: ciências agrárias e ambientais, Curitiba, v.3, n.1, p. 11-18, jan./mar. 2005.

LIMA JUNIOR, J. A.; SILVA, A. L. P. Study of the salinisation process to indicate measures to prevent saline soils. Enciclopédia Biosfera, Centro Científico Conhecer - Goiânia, 2010.

LINHARES, P. C. F. Phytomass production and macronutrient content of jitirana at different phenological stages. Caatinga (Mossoró, Brazil), v.21, n.4, p.72-78, October/December 2008.

MACÊDO, L. de S. Salinity in irrigated areas. João Pessoa: EMEPA - PB, 1988. 11 p. (EMEPA. Comunicado Técnico, 38).

MASCARENHAS, J. C., BELTRÃO, B. A., JUNIOR, L. C. S., MORAIS, DE F., MENDES, V. A., MIRANDA, J. L. DE F. Project to register groundwater supply sources in the state of Paraíba. Diagnosis of the municipality of Sousa, state of Paraíba. CPRM/PRODEEM   , Recife   , 2005.   Available at   : http://www.cprm.gov.br/rehi/atlas/paraiba/relatorios/SOUS206.pdf. Accessed on> 28 April 2014.

MEDEIROS, J. F. de.; NASCIMENTO, I. B. do. & GHEYI, H. R. Soil-water-plant management in areas affected by salts. In: GHEYI, H. R; DIAS, N. da S. & LACERDA, C. F. de. eds. Salinity management in agriculture: Basic and applied studies. INCTSal, Fortaleza - CE, 472p., 2010.

MELO, R. M.; BARROS, M. de F. C.; SANTOS, P. M. dos; ROLIM, M. M. Correction of saline-sodic soils by the application of mineral gypsum. Revista Brasileira de Engenharia Agrícola e Ambiental, v.12, n. 4, p. 376-380, 2008. In: Santos, M. A. dos, Recovery of saline-sodic soil by phytoremediation with Atriplex nummularia or application of gypsum. Recife, 2012. 13-14 p. Master's dissertation.

METZNER, A. F.; CENTURION, J. F.; MARCHIORI JÚNIOR, M. Relationship between degree of flocculation and soil attribute. In: CONGRESSO BRASILEIRO DE CIÊNCIA DO SOLO, 29, 2003. Ribeirão Preto-SP, ANAIS... Botucatu, 2003. CD.

MIRANDA, M. F. A. Diagnosis and recovery of soils affected by salts in an irrigated perimeter in the sertão of Pernambuco. Recife, 2013. 102p

MIYAMOTO, S.; GLENN, E. P.; SINGH, N. T. Utilisation of halophytic plants for fodder production with brackish water in subtropic deserts. In: Squires, V. R.; Ayoub, A. T. (ed) Halophyte as a resource for livestock and for rehabilitation of degraded land Proseedings. Amsterdam: Kluwer Academic, 1994. p.43-75.

MIYAMOTO, S.; GLENN, E.P. & OLSEN, M.W. Growth, water use and salt uptake of four halophytes irrigated with highly saline water. J. Arid Environ., 32:141-159, 1996. Master's thesis

MONTENEGRO, A.A.A. & MONTENEGRO, S.M.G. Sustainable utilisation of alluvial aquifers in the semi-arid region. In: CABRAL, J.S.P.; FERREIRA, J.P.C.L.; MONTENEGRO, S.M.G.L. & COSTA,W.D. Água subterrânea: Aquíferos costeiros e aluviões, vulnerabilidade e aproveitamento. Special topics in water resources. Recife, Federal University of Pernambuco, 2004. v.4. 447p. Master's dissertation.

MOTA, F. O. B.; OLIVEIRA, J. B. Mineralogy of soils with excess sodium in the state of Ceará. Revista Brasileira de Ciência do Solo, v. 23, n. 4, p. 799-806, 1999. Extract from the Dissertation: Santos, Monaliza Alves dos, Recovery of saline-sodic soil by phytoremediation with Atriplex nummularia or application of gypsum. Recife, 2012.

MUNNS, R. Comparative physiology of salt and water stress. Plant Cell Environment, v.25, n.5, p.659-662, 2002. Extract from the dissertation: Santos, Monaliza Alves dos, Recovery of

saline-sodic soil by phytoremediation with Atriplex nummularia or application of gypsum.Recife, 2012.

MUNNS, R. Genes and salt tolerance: bringing them together. New Phytologist, v. 167, n. 03, p. 645-663, 2005.

MUNNS, R.; TESTER, M. Mechanisms of salinity tolerance. Annual Plant Biologic, v. 59, p. 651-81, 2008.

MURAISHI, C.T.; LEAL, A.J.F.; LAZARINI, E.; RODRIGUES, L.R. & GOMES JUNIOR, F.G.G. Management of ground cover plant species and productivity of maize and soya in direct sowing. Acta Sci. Agron., 27:199-207p, 2005.

OLIVARES, A. The shrubs of the genus Atriplex and their importance as forage species. In: Actas del encuentro del estado de las investigaciones sobre manejo silvopastoral en Chile. Proyecto CONAF/PNUD/FAO. Talca, University of Talca, Department of Forestry. Forestry. P. 5-13, 1983.

OLIVEIRA, L. B. de; RIBEIRO, M. R.; FERREIRA, M. da G. de V. X.; LIMA, J. F. W. F. de & MARQUES, F. A. Pedological interferences applied to the irrigated perimeter of Custódia, PE. Pesquisa Agropecuária Brasileira, v.37, n.10, p.1477-1486, 2002.

PEREIRA, F. H. F.; ESPINULA NETO, D.; SOARES, D. C. & OLIVA, M. A. Gas exchange in tomato plants subjected to saline conditions. Horticultura Brasileira, v.22, n.2, 2005.

PIZARRO CABELLO, F. Drenage agricola y recuperación de suelos salinos. 2. ed. Madrid: Editorial Agrícola Espanhola S. A., 1985. 542 p.

PIZARRO, F. Drenaje agrícola y recuperación de suelos salinos. Madrid: Espanola. 521p., 1985.

PIZARRO, F. Drenaje agrícola y recuperación de suelos salinos. Madrid: Espanola. 1978.

PIZARRO CABELLO, F. Riegos localizados de alta frecuencia: goteo, microaspersión, exudación. 2. ed. Madrid: Ediciones Mundi-Prensa, 1996. 471p.

PORTO, E. R.; AMORIM, M. C. C. & SILVA JÚNIOR, L. G. A. Uso de rejeito de desalinização de água salobra para irrigação da erva-sal (Atriplex nummularia), Revista Brasileira de Engenharia Agrícola e Ambiental, v.5, p.111-114, 2001.

PORTO, E. R.; AMORIM, M. C. C. DE; DUTRA, M. T.; PAULINO, R. V.; BRITO, L. T. L.; MATOS, A. N. B. Yield of Atriplex nummularia irrigated with effluent from tilapia farming in water desalination rejects. Revista Brasileira de Engenharia Agrícola e Ambiental, v. 10, n. 1, p. 97-103, 2006.

PRADO, R.; NATALE, W. Changes in granulometry, degree of flocculation and chemical properties of a dystrophic Red Latosol, under no-tillage and reforestation. Acta Scientiarum: Agronomy, 25:45-52, 2003.

QADIR, M.; GHAFOOR, A. & MURTAZA, G. Use of salinesodic waters through phytoremediation of calcareous saline-sodic soils. Agric. Water Manag., 50:197-210, 2001. Master's thesis.

QADIR, M.; OSTER, J. D.; SCHUBERT, S.; NOBLE, A. D.; SAHRAWAT, K. L. Phytoremediation of sodic and saline-sodic soils. Advances in Agronomy, v. 96, p. 197-247, 2007. Extract from the Dissertation: Santos, Monaliza Alves dos, Recovery of saline-sodic soil by phytoremediation with Atriplex nummularia or application of gypsum. Recife, 2012.

QADIR, M.; OSTER, J.D.; SCHUBERT, S.; NOBLE, A.D. & SAHRAWAT, K.L. Phytoremediations of sodic and salinesodic soils. Adv. Agron., 96:197-247, 2007. Master's thesis.

RAIJ, B. van. Gypsum in agriculture. Campinas. Agronomic Institute, 2008. 233p.

RIBEIRO, M. R. Origin and classification of soils affected by salts. In: GHEYI, R.H.; DIAS, N.S.; LACERDA, C.F. Manejo da salinidade na agricultura: Estudos básicos e aplicados. Fortaleza: INCT Sal, 2010. 472p. Extract from the Dissertation: Santos, Monaliza Alves dos, Recovery of saline-sodic soil by phytoremediation with Atriplex nummularia or application of gypsum. Recife, 2012.

RIBEIRO, M. R.; BARROS, M. F. C.; FREIRE, M. B. G. dos S. Chemistry of saline and sodic soils. In: MELO, V. F.; ALLEONI, L. R. Química e mineralogia do solo. Viçosa: Brazilian Society of Soil Science, v.2, p.449- 484, 2009. Extract from the Dissertation: Santos, Monaliza Alves dos, Recovery of saline-sodic soil by phytoremediation with Atriplex nummularia or application of gypsum. Recife, 2012.

RIBEIRO, M. R.; FREIRE. F. J. & MONTENEGRO, A. A. A. Halomorphic soils in Brazil: Occurrence, genesis, classification, use and sustainable management. In: CURI, N.; MARQUES, J. J.; GUILHERME, L. R. G. G.; et al. eds. Topics in Soil Science. Brazilian Society of Soil Science, 2003, 3, p. 165-208.

RICHARDS, L. A. Diagnosis and improvement of saline and alkali soils. Washington: U.S, Department of Agriculture, 160p. 1954.

RUIZ, H. A.; Sampaio, R. A.; Oliveira, M. de; Ferreira, P. A. Physical characteristics of saline-sodic soils submitted to leaching slope instalments. Journal of Soil Science and Plant Nutrition, v.6, p.1-12. 2006.

RUIZ, H. A.; SAMPAIO, R. A; OLIVEIRA, M. de ; VENEGAS, V. H. A. Chemical characteristics of saline-sodic soils subjected to leaching rate instalments. Pesquisa Agropecuária Brasileira, v.39, n.11, p.1119-1126, 2004. Extract from the Dissertation: Santos, Monaliza Alves dos, Recovery of saline-sodic soil by phytoremediation with Atriplex nummularia or application of gypsum. Recife, 2012.

SDB, Handout on recovery of degraded areas, UFRRJ, 2015. Available at:

http://r1.ufrrj.br/cfar/d/download/Apostila%20de%20areas%20degradadas.pdf. Accessed October 2015.

SILVA, E. N. da; SILVEIRA, J. A. G.; FERNANDES, C. R. R.; DUTRA, A. T. B.; ARAGÃ, R. M. de. Ion accumulation and growth of jatropha under different salinity levels. Revista Ciência Agronômica, v. 40, n. 2, p.240-246, 2009.

SILVA, N.M. da; CARVALHO, L.H.; HIROCE, R. & KONDO, J.I. Response of cotton to the application of lime and potassium chloride. Bragantia, v.43, n.2, p.643-658, 1984.
Brazilian Soil Classification System. 2. ed. - Rio de Janeiro : EMBRAPA, 2006.

Brazilian Soil Classification System. SANTOS, H. G. - 3 ed. Brasília, DF: Embrapa, 2013.

SORATO, R. P.; CRUSCIOL, C. A. C.; MELLO, F. F. de C. Production components and productivity of rice and bean cultivars as a function of lime and gypsum applied to the soil surface. Bragantia, v. 69, n. 4, p.965-974, 2010.

SOUSA, D. M. G.; LOBATO, E.; REIN, T. A. Use of agricultural gypsum in Cerrado soils. Planaltina: EMBRAPA-CPAC, 1996. 20 p. (EMBRAPA-CPAC. Circular Técnica, 32).

SOUZA, Edivan Rodrigues. Phytoremediation of saline sodic fluvial neosol from Pernambuco with Atriplex nummularia., p.9. 2010 (Doctoral Thesis in Agricultural Engineering. Federal Rural University of Pernambuco. Department of Rural Technology. Recife. 2010. Available: http://200.17.137.108/tde_busca/arquivo.php?codArquivo=843, accessed July 2015.

SOUZA, E. R. de; FREIRE, M. B. G. DOS S; NASCIMENTO, C. W. A. do; MONTENEGRO, A. A. DE A. FREIRE, F. J.; MELO, H. F. de. Phytoextraction of salts by Atriplex nummularia lindl. under water stress in sodic saline soil. Revista Brasileira Engenharia Agrícola e Ambiental, v.15, n. 5, p.477-483, 2011.

SQUIRES, V.R. & AYOUB, A. Halophytes as a resource for livestock and for rehabilitation of degraded land. Dordrecht, Kluwer Academic Publishers, 1994. 481p. Master's dissertation.

SUASSUNA, J. & AUDRY, P. Salinity statistics of irrigation water in the Brazilian semi-arid Northeast. IN: SBPC Annual Meeting, 45th, 1993, (Recife). Proceedings. Recife: SBPC, 1993. p. 53-72. Doctoral thesis.

SUHAYDA, C. G.; Yin, L.; Redmann, R. E.; Li, J. Gypsum amendment improves native grass establishment on saline-alkali sol is in northeast China. Soil Use and Management, v.13, p.43-47, 1997.

TAVARES FILHO, A. N. Levels of need for gypsum on the physical-chemical characteristics and correction of saline-sodic soils in the irrigated perimeter of Ibimirim -PE. 81 p. 2010 (Dissertation - Master's Degree in Agricultural Engineering. Federal Rural University of Pernambuco. Department of Rural Technology. Recife. 2010. Available at:<http://www.pgea.ufrpe.br/downloads/dissertacoes/AntonioNovais.pdf> Accessed on: 11 May 2015.

TEIXEIRA, A.R.N.; PINTO-RICARDO, C.P. Fotossíntese. Coleção formação universitária. (Didática, Eds.) Lisbon, Portugal. 201-220 p, 1983.

UNITED STATES SALINITY LABORATORY - USSL Staff Diagnosis and improvement of saline and alkali soils. Washington DC: US Department of Agriculture, 1954. 160 p. (USDA Agricultural Handbook, 60).

YOKOI, S.; BRESSAN, R. A.; HASEGAWA, P. M. Salt stress tolerance of plants. JIRCAS Working Report, p. 25-33, 2002.

ZAMBROSI, F.C.B. et al. Nutrient concentration in soil water extracts and soybean nutrition in response to lime and gypsum applications to an acid Oxisol under no-till system. Springer Science+Business Media B.V, v.79, p.169-179, 2007.

ZHU, J. K. Plant salt tolerance. Trends in Plant Science, v. 6, p. 66-71, 2001. Extract from the Dissertation: Santos, Monaliza Alves dos, Recovery of saline-sodic soil by phytoremediation with Atriplex nummularia or application of gypsum. Recife, 2012.

ZONTA. E.; BRASIL, F.; GOI, S.R. & ROSA, M.M.T. The root system and its interactions with the soil environment. In: FERNANDES, M.S., ed. Nutrição mineral de plantas. Viçosa, MG, Sociedade Brasileira de Ciência do Solo, 2006. p.7-52.

I want morebooks!

Buy your books fast and straightforward online - at one of world's fastest growing online book stores! Environmentally sound due to Print-on-Demand technologies.

Buy your books online at
www.morebooks.shop

Kaufen Sie Ihre Bücher schnell und unkompliziert online – auf einer der am schnellsten wachsenden Buchhandelsplattformen weltweit! Dank Print-On-Demand umwelt- und ressourcenschonend produzi ert.

Bücher schneller online kaufen
www.morebooks.shop

Printed by Books on Demand GmbH, Norderstedt / Germany